J. W Meader

The Merrimack River

Its source and its tributaries. Embracing a history of manufactures, and of the towns along its course; their geography, topography, and products, with a description of the magnificent natural scenery about its upper waters

J. W Meader

The Merrimack River
Its source and its tributaries. Embracing a history of manufactures, and of the towns along its course; their geography, topography, and products, with a description of the magnificent natural scenery about its upper waters

ISBN/EAN: 9783337301507

Printed in Europe, USA, Canada, Australia, Japan

Cover: Foto ©berggeist007 / pixelio.de

More available books at **www.hansebooks.com**

THE
MERRIMACK RIVER;

ITS

SOURCE AND ITS TRIBUTARIES.

EMBRACING

A HISTORY OF MANUFACTURES, AND OF THE TOWNS ALONG ITS COURSE;
THEIR GEOGRAPHY, TOPOGRAPHY, AND PRODUCTS, WITH A DE-
SCRIPTION OF THE MAGNIFICENT NATURAL SCENERY
ABOUT ITS UPPER WATERS.

BY

J. W. MEADER.

BOSTON:
PUBLISHED BY B. B. RUSSELL, 55 CORNHILL.
1869.

TO

D. W. SMITH,

A SON OF THE MERRIMACK VALLEY, AND,

THROUGH MANY YEARS,

A True and Faithful Friend,

THIS VOLUME

Is Affectionately Inscribed.

PREFACE.

NEW HAMPSHIRE and Massachusetts naturally entertain a commendable and worthy pride in the joint possession of the magnificent Merrimack River, and properly appreciate the wonderful beauty of its scenery, its stupendous hydraulic power, and its incalculable value and importance, not alone to these two States, but to all New England, the nation, and to the world. As a great natural feature the Merrimack, it is believed, surpasses all others in the harmonious blending of the useful and the beautiful, and the facts assembled on these pages are confidently expected to warrant and justify this view. The source of the Merrimack being more than six thousand feet above the level of the sea, — much higher than that of the Connecticut, while it is only half the length of the latter stream, — it is, from its source to its mouth, literally, a vast system of mill-privileges with excellent water-power, material and conveniences for dams, and an ample and unfailing supply of water. The amount of manufacturing along this stream is not equalled by that of any other river in the world, while even yet many of its best mill-sites are unimproved. Having been for many years familiar with all the territory which supplies the Merrimack and its branches, it has been a matter of surprise that some competent person has not, long since, given the subject the attention it merits. It was not, however, until eminent oculists prescribed open-air exercise for a malady more frightful, inconvenient, and uncomfortable than painful, when the atmospheric purity of this region suggested it as a proper resort, that the idea occurred to collect some facts and place them before the public myself. Accordingly, without experience or pretension, the result is here presented; and while it is believed some of the matter is new, it is hoped much of it will prove interesting, and aid in extending a more thorough knowledge of this important subject. No literary merit is claimed, and whatever is open to criticism the author trusts may be kindly considered. Accuracy, both in the aggregate and detail, has been sought, and, if not always attained, may be attributed to erroneous but conscientious deductions, or to incorrect information. The hydraulic capacity of the Merrimack is only fully known to the most experienced practical engineers, and though its source is theoretically known to many, very few, through the great wild valley of the upper waters, have penetrated to it and inspected it personally. Indeed, the number is so limited as to be, probably, more than told on the digits. The great aorta and its branches have been traversed, and all its sections of especial interest received personal inspection; still it is to be regretted that many letters have failed to produce a response, which militates, to some extent, against individuals and interests as well as the completeness of the work.

Historical and geographical sketches have been made (the latter principally from personal observation) of towns on the river and its tributaries, not, however, as complete as could be desired. Biography has been only incidentally and briefly essayed; in all cases without prejudice, in the light of facts and the spirit of truth.

The design and object of this work have been to give a more particular account and description of the Merrimack River and its branches, together with the territory drained, of its capacity for manufacturing and mechanical uses, the history, geography and topography, with something of its matchless and unequalled natural scenery. It is true the sketch of the route from Lake Winnipesaukee to Mount Washington is not strictly a legitimate portion of the section under consideration, but it is uniformly traversed by all tourists through the valley of the Merrimack, — therefore ignoring it, would be like omitting the principal character of a play.

The different kinds of game which abound in the great forests bordering the upper waters of the Merrimack and its northern affluents, the varieties of fish worthy the angler's consideration, the various as well as best methods of taking them, have received considerable attention. This feature, it is thought, may be of marked and especial interest to a multitude of gentlemen, experts, amateurs, and tyros, who yearly resort to this region to enjoy the scenery and indulge in field sports.

To those at home or abroad who retain a natural and creditable State pride; to those who have a present and prospective pecuniary interest; to that numerous and valuable class who have contributed most to the immense business upon the Merrimack, mechanics, artisans, and operatives; to all who would know the beauties and the resources of the territory of this marvellous and most resplendent stream, this volume is offered, with the earnest desire that it may be of some value and aid to them in the effort to obtain such information.

CONTENTS.

CHAPTER I.

New Hampshire. — Voyage of Columbus. — The Cabots. — Capt. John Smith. — Pierre du Guast. — De Champlain. — Discovery of the Merrimack. — Mason and Gorges. — Laconia. — The first Settlement of New Hampshire. — Wheelwright. — The Puritans. — Witchcraft. — Rapid Settlement of New Hampshire 9

CHAPTER II.

The Pemigewasset. — Mountains in Summer. — In Winter. — Source of the Merrimack. — Method of taking Salmon and Trout. — Franconia Notch. — Echo Lake. — The Flume. — The "Old Man of the Mountain" 40

CHAPTER III.

Franconia. — Lincoln. — Woodstock. — Thornton. — Game and Gunning. — Campton. — Holderness. — Samuel Livermore. — Plymouth. — Rumney. — Wentworth. — The Moosiaukes. — Bridgewater. — Squam Lakes and River. — New Hampton. — Bristol. — Newfound Lake and River. — Hill. — Smith's River. — Orange — Andover. — Salisbury. — Daniel Webster. — Kearsarge. — Boscawen. — Franklin . . 65

CHAPTER IV.

The Forks. — Winnipesaukee Lake and River. — Pickerel Fishing. — The Wiers. — Laconia. — Capt. Lovewell. — Centre Harbor. — Moultonborough. — Red Hill. — Sandwich 98

CHAPTER V.

Tamworth. — Quakers. — Albany. — Chocorua. — Madison. — Mines. — Conway. — The Notch. — The Willey Family. — White Mountain Railroad 121

CHAPTER VI.

Gilford. — Alton. — Wolfboro'. — Tuftonborough. — Meredith. — Sanbornton. — Northfield. — Canterbury. — Shakers. — Pembroke. — Suncook River. — Gilmanton. — Barnstead. — Pittsfield. — Epsom. — Allenstown. — Contoocook River. — Hillsboro'. — Gov. Pierce. — Henniker. — Washington, etc. — Hopkinton . . . 147

CHAPTER VII.

Concord. — The Pennacooks. — First Settlement. — State Institutions. — Ex-President Pierce. — Isaac Hill. — Count Rumford. — Low. — Hooksett 167

CHAPTER VIII.

Amoskeag Falls. — Indians. — Fisheries. — Manchester. — History of its Manufactures. — The Starks. — City Institutions. — Samuel Blodgett 185

CHAPTER IX.

Cohas River. — Massabesic. — Londonderry. — Scotch-Irish Settlement. — Distinguished Men. — Derry. — Piscataquog River. — Francestown. — Weare. — Goffstown. — Bedford. — Souhegan River and Towns along its Course. — Litchfield. — Reed's Island. — Hudson. — Nashua River and the Towns watered by it. — Dunstable. — The Pequauket War. — Nashua. — Tyngsboro'. — Chelmsford. — Stony Brook. — Dracut. — Beaver River. — John Nesmith 217

CHAPTER X.

Pawtucket Falls. — Indians. — Canada. — Lowell. — History of Manufactures on the Merrimack. — The Concord River. — Its History. — History of its Manufactures . 242

CHAPTER XI.

Pentucket Navigation Company. — Nicholas G. Norcross. — Andover. — Methuen. — The Spicket. — Lawrence. — History of its Manufactures. — Bradford. — Little River. — Haverhill. — The Pow-Pow. — Amesbury. — History of its Manufactures. — Newbury. — Salisbury. — Newburyport. — Plum Island. — Seabrook. — Conclusion 284

THE MERRIMACK RIVER;

ITS SOURCE AND ITS TRIBUTARIES.

CHAPTER I.

New Hampshire. — Voyage of Columbus. — The Cabots. — Capt. John Smith. — Pierre du Guast. — De Champlain. — Discovery of the Merrimack. — Mason and Gorges. — Laconia. — The First Settlement of New Hampshire. — Wheelwright. — The Puritans. — Witchcraft. — Rapid Settlement of New Hampshire.

THE State of New Hampshire, limited in territorial dimensions to a mere patch or speck on the map of the continent, with a climate rigorous and inhospitable, a rugged, sterile, and comparatively unproductive soil, possesses, notwithstanding, these physical disabilities, and many others, a rare combination of those necessary and peculiar elements which alone secure a nation's greatness, and exhibit a community intelligent, prosperous, and happy.

The history of New Hampshire extends over a period of more than two centuries, and though it embraces a broad and varied record, it is still replete with patriotic works and fearless daring, and noble achievements. Planted in a wilderness, in the very midst of wild and ferocious beasts, and still more intractable, implacable, and savage men, her colonists had no other alternative but to push or be pushed. With them it was to be a great success or a great failure. With such men as the indomitable pioneers of New Hampshire, — men, who in company with the early navigators, had led the van along the trackless and hitherto unknown seas, or bared their breasts to the pitiless pelting of the lead and iron hail of internecine strife; or, better still, had wrung from mother earth a sustenance, and a tribute to their industry and skill with the plough, the spade, and the sickle, — with such men as these there could be no such word as fail; the result was inevitable, and with broad shoulders, strong arms, stout hearts, and masculine intellects, they have left a record to which their State, their country, and their race may point with just and worthy pride.

To err is human; therefore perfection in any direction may not be claimed; still the bright galaxy of names which adorn and embellish each era of her history most clearly proves that her boast, her pride, or her claim is certainly, to say the least, well founded.

Although, as it has been observed, the State is one of the smallest in territory, in resources, and in natural advantages, still it is claimed that she is unsurpassed, perhaps unequalled, by any of her sister states, — always excepting the Old Dominion, the mother of states and of statesmen, — in the number and character of her illustrious men. Her statesmen have been conspicuous; her jurists learned, pure, dignified, and famous; her soldiers among the bravest and foremost captains of every age; her mechanics displaying a versatility and skill unsurpassed; the sturdy, undaunted, and persevering tiller of the soil, by his genius, industry, and judgment, wrenching a reputation as enduring as the granite base of the land he cultivates. Even the women have displayed in a marked manner those ennobling traits which recognize them as fit companions and educators of such a race of men. The noble-hearted mothers, wives, sisters, and daughters, affectionate, unselfish, devoted, and self-sacrificing, softened down the rough angles so indigenous to the isolated pioneers, especially such as the settlers of New Hampshire, who confronted a primeval wilderness filled with wild beasts and untamed savages, and had at the same time a covetous and encroaching neighbor in Massachusetts, constantly putting forward unfounded claims to the territory to impose taxation and entire jurisdiction; and although these claims, unjust and arrogant, were successfully resisted, still they served to stir up and keep alive an unnecessary and annoying irritation, and bred in the minds of the weak colonists a natural and not unavailing determination to realize, cost what it would, those indescribable emotions that come of independence, which they felt they were entitled to, and were determined to achieve.

Actuated by a spirit and resolution which showed what manner of men they were, they pushed forward without any deflection, encountering and overcoming obstacles of such magnitude as would seem to appall the stoutest heart, determined, regardless of expense, in trials, privations, and hardships, to provide a sustenance, establish a home, and erect a sovereign State. What these self-reliant and independent Scotch-Irish, and English, and their descendants have achieved, history and geography fully explain, for they have left their impress on every page of the one and every parallel of the other.

It has been said in contempt or derision by some thoughtless, envious, or malicious person, that New Hampshire was a good State to emigrate from. This would-be aspersion is rather a credit than otherwise, as it simply signifies that the starving Josephs of other States less productive may confidently journey hither for a generous supply of wholesome intellectual food. When a native of good ability and intelligence, thoroughly imbued with the spirit of the State, emigrates, carrying with him and adhering to those sound principles which his State and his education have endowed him with, he will generally, in any sphere of life, ultimately achieve success, and in as far as deserved fame and prosperity attend them, they will, of course, reflect honor and credit on the mother State.

This as a rule is proved and demonstrated by the slur referred to. Taking an observation across the continent, it will be seen that the leading men of many States, men of mark and of note, justly claim New Hampshire as their fatherland. Statesmen, lawgivers, orators, journalists, merchants, conspicuous and successful, are seen on every hand, an ornament and a blessing to the community of their adoption, and a source of pride to those who have furnished a member who is equally as necessary to them as ornamental.

Scarcely a town in the State but boasts of some famous native who now occupies an exalted position in another State. Not unfrequently many of these wandering sons, when the care and labor of the busy outside world will allow, revisit their early homes to renew and refresh their thirsty spirits, and drink in the old familiar scenes of childhood, and live over again for a brief period those old affections of by-gone times: cooling the heated brain and relaxing the overwrought nerves in the quiet retreat, fanned by the same refreshing breezes, shaded by the same grand old trees ; to take a draught from the same old oaken bucket : and, looking through the wide rooms, stop before the quaint old portraits of departed ancestors ; or walk through the silent aisles of that peaceful and quiet neighborhood where the bones of an honored ancestry repose, to decipher again for the twentieth time perhaps, with humid eyes, the hieroglyphics wrought long years before by the hand of affection, which chronicle the deeds and worth of the deceased; while the majestic elms afford a calm and grateful shade, and the wild birds flit from branch to branch without fear, or carol plaintive and joyous melodies as if to do homage to the worth of the departed, and the modest little wild flower dots the vernal spread upon their lowly bed, and offers the humble tribute of a grateful fragrance to their memory.

Honored by the land of their adoption often with great and all-important interests in their keeping, severely taxed to preserve the trust reposed in them, the emigrant from this to other States still finds time for an occasional pilgrimage to his native land.

Of all the countless numbers of those who have emigrated to other States and mingled in other scenes, how few there are who have not conferred honor upon the land of their nativity! Daniel Webster may be cited as a conspicuous example of the honor reflected upon the State by her absent sons. Nor has emigration depleted the moral and intellectual wealth of the State, as is proved by the long line of great men, past and present, illustrious in every sphere of life and usefulness.

Deservedly great as has been the fame of her public men, her mechanics and artisans are equally celebrated; her monster brick palaces, which cluster around each available waterfall, exhibit an architectural ability equal to the best; her endless variety of implements and articles of wood manufacture display an ingenuity and finish unsurpassed; her locomotive engines roll on every gauge across the continent, bearing a precious freight of human souls to the fertile prairies that stretch away to the Rocky Mountains, returning with their monster trains of that indispensable merchandise which feeds and moves the arm of the thousands of mechanics at the East; her fire annihilator, the admiration of all, worked by the untiring and enduring muscle of steam and iron, the most efficient guardian of life and property, has raised its protecting ægis, not only in all the cities throughout the Union, but has carried the same invaluable quality, as well as the fame of our mechanics, across the Atlantic, and even to those remote lands beyond the great, tranquil sea.

The machinery, of every kind and for every purpose, produced by the industry and skill of her mechanics, is a marvel of perfection: whether designed for great strength or extreme delicacy, its model, its finish, its intricacy, its adjustment, its adaptation to and its efficiency in the service for which it was intended, cannot be excelled. The manufacture of every variety of cloth is known in the markets of the world, and whatever fabric is required for necessity, convenience, comfort, or ornament is produced.

From east to west, from north to south, the products of her factories are indispensable, and are turned out at prices so reasonable as to be within the reach of all. The successful prosecution of this great interest has not only been a benefit to our own land, but is a substantial blessing to the human race. The enterprise which has produced these wares invades all lands, and barters them to the uttermost ends of the earth, so that the naked and unsightly barbarian, the Hindoo and the Hottentot, as well as the Christian, may make a more presentable appearance in a full and not unbecoming costume of Manchester matchless cotton fabrics.

Thus it will be seen that in intellect, patriotism, the mechanic arts, and agriculture, in all the appliances which tend to human progress, the people of this rock-ribbed commonwealth maintain unimpaired the high character of their glorious ancestry. It may not be disputed, nay, it cannot be denied, that all communities in all time have had their faults and follies, and the State of New Hampshire is certainly no exception; still it is confidently believed that no community of equal numbers exhibits greater virtues or less vices, greater capacity in the science of government, in the art of war, or in the peaceful pursuits which adorn and dignify a people. Nor are the people alone the only object worthy of reference. Here nature presents herself in one of her most romantic moods. The surface of the State is undulating in all its length and breadth, while at the north many huge mountains rise to a dizzy height and are piled up in inextricable confusion, and lakes, rivers, and mountain torrents diversify the entire surface. This section has often been called the Switzerland of America, from its real or fancied resemblance to that romantic land. However this may be, there is a peculiar grandeur and sublimity about this region, as well as other distinguishing features, such as may be termed wonderful freaks of nature, which destroy the parallel, and leave this territory pre-eminently and incomparably sublime. That this is not overdrawn is attested by the thousands who annually visit here; and even the polished and scholarly Everett has touched it with the magic of his masterly eloquence.

Is it surprising, then, that a people reared amid such magnificent surroundings, and familiar with nature in its noblest aspect, — is it strange that a people thus situated should carve out for themselves a record which commands the attention and respect of mankind?

More than two hundred years ago the State of New Hampshire was settled by several slender colonies; some by the authority of crown patents, others under the king's grant, — sometimes made as a reward for discovery, — or under a semblance of purchase from the Indians, who, being the occupants, were, consequently, the unquestionable proprietors. The colonies were few and feeble, and, from various causes, this condition remained unchanged for many years. Owing to the severity of the climate and the unmitigated hostility of the Indians, the settlements maintained a sickly existence or dwindled away, and, finally disheartened and discouraged, broke up, and either joined their more fortunate neighbors, or returned to Europe with some of the many exploring expeditions which were constantly visiting the eastern coast. The progress of the colonies was slow; less determined people would have succumbed to the almost insurmountable obstacles which they encountered. As it was, it seemed that their holding was very precarious, and it was apparently doubtful if they could maintain a foothold on the soil. But fail, as shown by the result, was not in their vocabulary, and ere long the forest began to melt away before their vigorous blows, and it soon became apparent that it was only a question of time when the Indian and his title should both be extinguished.

While the savages were much the most numerous, equally brave, cunning, sagacious, and active, their training, habits, and implements of warfare were so much more primitive as left an immense disparity between the belligerents. The red man's wants were few and easily supplied, the field, the forest, and the flood affording all his necessaries, comforts, and luxuries of life, — the former contributing so little, however, as to be scarcely worth taking into account. His ability to prosecute a successful offensive warfare was limited to an ambuscade, or an unexpected midnight attack, while his defensive power was absolutely nothing; for, though the bow and arrow and tomahawk were really efficient in supplying food, the longer range and more deadly musket, sending forth its lightning and thunder, cutting down the chief, the brave, the squaw, and the pappoose indiscriminately, carried terror and dismay through their decimated and demoralized ranks. His villages being unprotected became an easy prey to the irresistible pale-face, and were plundered and burned without pity, and his only safety was to place a respectful distance between his habitation and his enemy.

Of the many voyagers to the Western or New World, as it was called, Columbus appears to have been the most unfortunate, as well as first and greatest. With a genius towering above all navigators and explorers who had hitherto been renowned and rewarded, and which was almost beyond the comprehension of princes, and an ambition lofty, unconquerable, irrepressible, and commendable, he bent his great mind and untiring energies to the task of securing, organizing, and fitting out an expedition which has, by universal consent, placed his name in the highest niche reserved for the mighty travellers of the sea in the temple of fame. Having prosecuted the most successful experiment ever attempted, and brought the problem of a Western World to a positive solution, and added at the same time its boundless area and fabulous wealth of undeveloped resources to the possessions of his ungrateful sovereign, he was permitted to languish in obscurity, and died broken-hearted in his fifty-ninth year. His patroness and only reliable friend, Isabella, having preceded him to the tomb, his compeers and rivals, actuated by a spirit of unworthy and ungenerous selfishness, which seems to have been connived at, or at least not restrained, by the king, stripped him as far as they could of the merit and the advantages resulting from the most magnificent achievement which human genius had ever devised and brought to successful issue. Thus Columbus proved to be no exception to the general rule, that great discoverers and inventors, the very few who possess originality of mind and genius, fail to realize the benefits of their achievements, which are usually secured by seedy and needy adventurers, — soldiers of fortune, whose shrewdness and enterprise is of that discreditable character which manages to appropriate the products of other greater and better minds.

However, though Columbus was shorn of the immediate fruits of his great enterprise by unbecoming envy, jealousy, malice, treachery, and gross ingratitude, impartial history will render all the justice possible to one of the most intrepid, ambitious, and enterprising navigators.

When the success of Columbus became known at home the Te Deum was sung,—the usual anthem for the most important of Spanish discoveries,—and the news was everywhere received with the wildest demonstrations of joy. The brilliant success of Spain inflamed the ambition and cupidity of England, and before the end of the second year she had fitted out, organized, and equipped a powerful expedition for exploration and discovery in the New World. The command and management of this important expedition was placed in the hands of John Cabot and his son Sebastian, gentlemen reputed to be possessed of great maritime skill. Sailing from Bristol, and pursuing a westward course, they discovered a large island which the sailors called Newfoundland, or Prima Vista; continuing their course they discovered other smaller islands, which they named, and soon reached the coast of Labrador, being the first Europeans who had ever actually seen the American continent.

Cabot explored the eastern coast as far south as Virginia, laying claim to the continent in the name of his sovereign.

The same spirit of detraction and envy which had been aroused by the success of Columbus, pursued the Cabots, and the consequence of this expedition, successful in all respects, was little else than to make their names historic to the extent of the improvement they had made on the previous voyage of Columbus. From this time forth expeditions were constantly sailing to the West for various purposes, but generally for trade and fishing.

Henry VII., under whose auspices the Cabot expedition was instituted, fitted out several others, for the purpose of finding a shorter route to the East Indies, which proved disastrous, and the attempt was abandoned. But the fisheries were prosecuted extensively by the maritime powers, as well as commerce with the natives of the islands and the coast along the north and east of New Hampshire for furs, and as far south as the Carolinas.

Though it is known that many vessels must have traded along the coast of New Hampshire and Maine, it is a remarkable fact that for a hundred years after the discovery of America no white man stepped foot upon its shores, or even saw it.

During the long intervening period between the discovery of America by Columbus and the discovery of New Hampshire by Capt. John Smith, the commerce of France, England, and Holland with the New World was considerably increased and the area of their navigation greatly enlarged. Many companies were formed in these countries, composed of gentlemen of wealth, enterprising merchants, and persons of a speculating turn of mind, who had in their employ the most skilful and daring officers to be found in the merchant marine. The ships of these companies visited the islands and explored the entire east coast of North America, bartering with the natives for furs, peltries, fish, sassafras, gums, and a great variety of other articles, for which they generally exacted knives and other weapons and implements made of steel. Though these companies generally failed to realize any remunerative gains for themselves, they opened and established a trade so extensive and varied, that ultimately, very much of the thrift and opulence of those countries grew out of it, and actually depended upon it. At the same time the American fisheries were prosecuted by these countries with great energy and considerable profit,—France alone having employed on the Grand Banks (1575–80), as many as one hundred and fifty sail of fishermen. While these countries were thus laying a firm foundation for a healthy and steady, rather than spasmodic growth, preparing the way to transplant a vigorous and enterprising race to the fertile lands of the New World who should compensate them by a development of its boundless resources, Spain turned her attention to the more genial climes of Mexico and South America. The history of her conquest of these countries is written in cruelty, cupidity, rapacity, and blood. The tracks of Pizarro, Cortez, and Almagro and their compeers exhibited more of desolation to the effeminate and terror-stricken natives than the hurricanes, roaring deluge, or consuming fire. Mad with the passion for gold, they swept like a whirlwind through half-civilized hordes of natives, and gathered their booty amidst the ruins of cities. Their love of riches was only equalled by their contempt for industry; they hurried the Indians in crowds to the mountains and forced them with merciless rigor to the fatal toil of the mines. Inflexible pride, determined valor, and deliberate atrocity marked their whole career of conquest and oppression. It was a union of avarice, fanaticism, and chivalry.

The native emperors, incas, and people were involved in common destruction; never were courage, fortitude, and valor devoted to more sordid, unjust, and barbarous ends; never was genius more powerfully employed to scourge mankind. "The paganism of the natives had allured to the invasion a few of the old bigots of Spain, and it is sadly instructive to mark these champions of the cross, trampling, in the name of religion, upon the most sacred rights, and giving glory to God amidst the destruction of life and the desolation of empires."

The founding of colonies generally comprehends, or at least involves, the expatriation if not extermination of the native inhabitants, if there be any, and though rapacity and flagrant cruelty may have, and undoubtedly did, mark the conduct of some of the early explorers and colonists of this country, still, unlike the experience of the Spaniard, the human havoc was not entirely one-sided; the native population found here, by the Europeans, were physically hardy, powerful, active, cunning, and warlike, and by their sagacious forecast, the primitive weapons, which had even come to be fatally efficient in their hands, had measurably given way to those which they had obtained by way of trade with inadvertent speculators, and which were much more convenient and effective for their protection, and for general warfare. Hence the North American Indian was under no compulsion to, and did not, remain passive until destroyed or subjugated, but often met his antagonist in deadly conflict under conditions regarded by himself as more than even, or, failing in this, he retired to the inaccessible wilds, where, to say the least, as his property was generally personal and portable, he was perfectly secure in his person and effects.

Notwithstanding the fact that many navigators were familiar with the north-east coast, and for many years trading and fishing had been engaged in by the maritime nations of Europe, and the most skilful, enterprising, and ambitious captains had been uniformly employed in this service, still there appears to be no record or even tradition of any European having touched or ever observed the coast of New Hampshire, until Capt. John Smith, sailing along the coast from Penobscot to Cape Cod, discovered, entered, and explored the fine harbor of Piscataqua, and caused a chart to be made of the territory, as far as he was able to ascertain, of the coast of New Hampshire and of the adjacent waters, which on his return he presented to the prince.

Capt. John Smith appears to have been even from childhood one of those erratic individuals who, moving without rudder or compass that can be observed, without visible aim or purpose, yet succeed in the accomplishment of some beneficent object by the very vicissitudes of fortune. A rollicking, vigorous, and ungoverned recklessness resulting in benefit to the world seems to prove these singular characteristics to have been in his case but the eccentricity of a universal genius.

At the very door of his teens, impatient of the slow process of acquiring knowledge at school, and disgusted with his books, he peddled them out, using the proceeds to secure some slender means for the accomplishment of an idea he had in contemplation even at that early age, — a clandestine voyage to sea.

His subsequent career the world knows by heart, and as long as the generous affection and unselfish devotion of Pocahontas shall be recognized as a conspicuous exhibition of untutored and spontaneous Christianity, it cannot be forgotten. Whether on an exploring expedition on the high seas, or planning the details of a colonial settlement, the same vigorous ideas and the same activity in putting in practice are everywhere apparent. He was the founder of colonies and the father of States. Did anything occur to derange the machinery of government, or the material prosperity of the colony, it appealed to Captain Smith as the only star of a last surviving hope to rescue it from impending anarchy, starvation, or extermination at the hands of a wily and ruthless savage foe. The story of his hair-breadth escapes, his trials, and triumphs is not new; but it is good and true; how he eluded the vigilance of the savages, successfully accomplished his objects, and saved the colonists, and, when finally made captive and sentenced to death, in his dire extremity, the fertility of resource never leaving him, he presented his executioner — that was to be — with a mariner's compass, pointing out to him the marvellous adjustment of the needle, its invariable attraction toward the north star; explained the vastness of the ocean, and the shape of the earth and its motion. The murderous design was suspended, and the superstitious Indians, overwhelmed by the magnitude of his wisdom and the incomparable superiority of his acquirements, granted him a reprieve until his case should be laid before Powhatan, the king, and his fate definitely settled.

Although by this cunning stratagem, he had delayed the decision of his fate, still he was not free; it is true he was held as an illustrious prisoner, and was regarded with superstitious veneration by his captors, and it was believed he could act as mediator to appease the wrath of the Great Spirit toward themselves.

Having several times escaped from the jaws of death, he lived to return home, and carried with him such accounts of the country he had explored as created the liveliest interest, and Charles complimented the country by calling it New England. Of Captain John Smith, one of his biographers says:—

"Whether we view him embarking for Italy with 'a rabble of pilgrims,' mounting the deadly breach at Regal, fighting hand to hand with the Turks in the armies of Austria, wandering in the deserts of Circassia, conducted a prisoner in the country of the Cambrian Tartars, passing over into Africa and visiting the court of Morocco, or surveying the wild coast of New Hampshire, he appears everywhere to be equally remarkable for his eccentric genius and his strange fortune."

A grant was made of that portion of the territory which had been christened New England, which was included between the fortieth and forty-eighth parallel north latitude, to a company consisting of forty distinguished gentlemen, some of whom had received the honor of knighthood. The financial and governmental management and control of this territory were to be perpetual in the hands of this company, and a majority vote of the shareholders in the corporation was to be a final and irrevocable decision on all subjects connected with its prosperity and its progress.

Thus the charter of New England was the foundation on which was based very many of the subsequent patents, grants, and charters which were made, and by which New England became ultimately settled. The views, plans, schemes, and ideas of these individuals were no doubt multifarious and varied, perhaps visionary. However that may be, the first influence that was felt, or which affected New England, was corporate power,—a mighty power which, as experience has proved, may be wielded with tremendous consequences for weal or woe. But this corporation, like all irresponsible powers, at length becoming an unendurable and odious monopoly, complaint was made that it was a serious grievance, and the charter was finally revoked.

Previous to the explorations of Captain Smith an expedition was organized at Bristol, which was put under the command of Captain Pring. It consisted of two small vessels, — the Speedwell and Discoverer, — and about fifty men, which sailed on the 10th of April, 1603, and sailed along the coast of Maine, as far as the Piscataqua River, which, it is said, he explored. He, however, seems to have confined his explorations principally to the coast of Maine, and thus this expedition seems to have thrown little or no light on the dark and unknown wilderness of New Hampshire, and contributed no information or knowledge concerning its inhabitants, topography, resources, or extent. The name of New England, which this territory has retained, is by no means the only appellation which it has received; among other names by which from time to time it has been known are New Spain, New France, New Holland, Acadia, etc.

In 1603, Pierre du Guast, Sieur de Monts, obtained an exclusive patent of the country called New France, from the fortieth to the forty-sixth parallel, under the name of Acadia; De Monts having been joined by De Champlain, a navigator of considerable ability, who had already come out with an expedition which sailed up the St. Lawrence, and selected Quebec as the site for a strong fortification, and learned much of the disposition, character, and numbers of the native population as well as something of the topography of the country. The expedition sailed from Havre, March, 1604, consisting of four vessels. Arriving in these waters the vessels parted company for various destinations and purposes. Champlain with others, in 1605, proceeded to explore the coast as far south as Cape Cod, with a view to locating settlements; but the natives appeared so numerous and unfriendly that they were deterred from completing this design.

De Champlain entered and surveyed the harbor at the mouth of the Piscataqua, and when near the Isles of Shoals he discovered some natives on the opposite shore. Wishing to obtain what information he could, he proceeded toward them to seek an interview, when they met him in a canoe; after making them presents of bread and some implements, which they received with manifestations of delight, he inquired of them concerning the lay of the land, the direction of the coast, and other matters; in reply, they sketched a rude diagram or chart of the coast, which proved to be extremely accurate. In the chart which they traced, they informed him there was a large bay to the westward, into which flowed a great and beautiful river. The next day De Champlain weighed anchor and sailed, determined to make the land he had seen far to the southward, and which is now known as Cape Ann; proceeding, he discovered Plum Island and the mouth of a magnificent river. This he called "Riviere du Guast, which in my judgment rises in the direction of the Iriquois, — a nation which is at open war with the mountaineers who are upon the great River St. Lawrence." De Champlain, it appears, based his opinion concerning the course of the river on what he was able to see of it, and his conjecture was erroneous, as, instead of its course being westward, it is northward. Thus was De Champlain the discoverer of the Merrimack River, although its existence was previously known, even under its present name, by the coast Indians and Europeans far to the eastward. Some accounts declare that De Champlain landed at the place of the interview with the natives, which was undoubtedly what is now known as Rye Beach. If this was the case, he was unquestionably the first European who ever set foot on the soil of New Hampshire; but the fact may be considered as still in doubt whether or not he actually landed.

It will be seen that the fame of this river had extended hundreds of leagues, and its distinguishing characteristics were chronicled by the aborigines in the name they gave it.

"This river was called Merrimack by the northern Indians. Merrimack means, doubtless, a place of strong current, from Merroh (strong) and awke (a place), the "M" being thrown in for the sake of sound. But by the Massachusetts Indians this river was called Menomack, from mena (an island) and awke (a place), meaning the island place, from the number of beautiful islands in this river."*

The Indians, it is said, in communications, messages, and conversation among themselves, used many descriptive expressions when speaking of the Merrimack River, such as "the bright, rapid water," "the beautiful river with the pebbly bottom," "the water

* Potter.

that comes from the high place," etc. etc. They were certainly great admirers of the grand and beautiful in nature, and the names which they bestowed upon localities, mountains, lakes, and rivers, exhibit incontestable proof of this, and in many instances even an adoration of the Great Spirit, to whom they naturally attributed the authorship of all things.

But these barbarians, like their more enlightened followers, looked upon this river more with an eye to business than to beauty; to its utility and capacity for augmenting their material prosperity than to its merely ornamental qualities. Thus even the name of this river had an unmistakable significance in it, as it described just such a river as they well knew that the cream of all the migratory tribes would most delight in, and where they were certain to crowd in countless myriads on their regular annual summer excursion from the great water to the foaming cascade-ripples, dark pools, and shaded reaches, where the numerous tributaries roll down from the mountains, and meandering through long stretches of forests, afford those cool retreats to which they were accustomed to resort for summering and spawning.

As the artisan, mechanic, manufacturer, and capitalist of modern times learned the adaptation and capacity of this river as a motive power, and congregated about its falls, — constructing dams to catch the water on its journey down, erecting monster mills with massive six-story brick walls, building populous cities and beautiful villages, fabricating unsurpassed cotton and woollen goods, machinery, and implements, — so did the red man of primitive days learn the peculiar fitness of this river for supplying his necessities and wants, equally as pressing and urgent as those of his more civilized successors; and here he also congregated around the same falls, constructing weirs to catch his food, erecting his smoking-poles and drying-flakes, building his little wigwam city, ornamenting his blankets and moccasins, growing his scanty supply of maize, pompions, gourds, and other varieties of squash, catching, curing and trafficking in salmon, shad, eels, and ale-wives; so that the Lowell, Manchester, and Laconia of the present day are but the Pawtucket, "Namoskeag" and Winnipesaukee of by-gone times,— a little exaggerated, to be sure, and differing somewhat in the kind, quantity, and value of their productions, still the same busy marts, crowded as now with a community — which, however, differed from that of the present day in every respect, except the single one of being industrious — at the time of the spring and fall migration. As has already been stated, the Merrimack River was widely known and famed among the Indians long before its discovery by Europeans, and, further, it was known everywhere among the natives by the name it still retains. It is true its European discoverers made a futile attempt at rechristening, but it is believed that even a legislative enactment changing its name would be ineffectual and unpopular, and that any new name would not survive the baptismal-day except in compulsory legal forms.

"The discovery of the Merrimack took place under the auspices of Henry IV., commonly called Henry le Grande (the great), whose reign forms one of the most brilliant eras in the annals of France. In 1603 Pierre du Gaust, Sieur de Monts, one of the ablest of the Huguenot chiefs, obtained a patent from this king creating him lieutenant-general and vice-admiral, and vesting in him the government of New France, which embraced all our eastern and middle States together with the Dominion of Canada. On the seventh of March, 1604, De Monts sailed from Havre with an expedition for colonizing Acadia, as these new dominions were called. He arrived on the sixth of April, and began at once the great work of exploration and settlement. While talking with the Indians on the banks of the River St. Lawrence, in the ensuing summer, he was told by them that there was a beautiful river lying far to the south, which they called the Merrimack.

"The following winter De Monts spent with his fellow-pioneers on the Island of St. Croix, in Passamaquoddy Bay, amid hardships as severe as those which sixteen years later beset the Pilgrims at Plymouth. On the eighteenth of June, 1605, in a bark of fifteen tons, having with him the Sieur de Champlain and several other French gentlemen, twenty sailors, and an Indian with his squaw, De Monts sailed from St. Croix, and standing in to the south examined the coast as far as Cape Cod. In the course of this cruise, on the seventeenth of July, 1605, he entered the bay on which the city of Newburyport now stands, and discovered the Merrimack at its mouth." The Sieur de Champlain, the faithful pilot of De Monts and chronicler of his voyages, has left a notice of this discovery in a work which ranks among the most romantic in the literature of the sea.

In closing this notice, Champlain says: —

"Moreover, there is in this bay a river of considerable magnitude, which we have called Gua's River. Thus De Monts named the Merrimack from himself, but the compliment was not accepted. Regardless of the name with which it was baptized by its discoverer, the Merrimack clung, with poetic justice, to the name which it received from the Indians long before the flag of the vice-admiral floated over Newburyport Bay. The visit of Admiral De Monts, like that of Capt. John Smith in 1614, was attended with no result. Other renowned names were yet to be inscribed on the list of the visitors of the Merrimack; but its song was the song of Tennyson's brook: —

'For men may come and men may go,
But I roll on forever.'"*

Among the members of the Plymouth Council there were two men named John Mason and Ferdinand Gorges. These two men had already acquired something of a reputation; but their history — that of Mason especially — is rightfully the property of New Hampshire. Equally enterprising, ambitious, and selfish, the minds of both were inflamed by the marvellous and incredible stories of the New World related by returned adventurers; and being thrown much into each other's society, they conceived the idea and put in execution a project for together obtaining a grant and colonizing in the New World. Capt. John Mason had been appointed governor of Newfoundland, and being imbued to some extent, doubtless, with the spirit of credulity, he had, from the island, cast a wishful and longing eye toward the mainland, even down to the more favored regions of New England, the recently discovered El Dorado, where it was thought that untold wealth might reward those who sought it there.

On his return to England, Capt. Mason obtained from the Council a grant which comprised all the territory from Naumkeag † River,

---

* Relations des Jesuits.

† This word Dr. Increase Mather declared to be of Hebrew origin. *Ergo*, the North American Indian was a descendant of that ancient race. He first writes the word "*Nahumkeik*," and then goes on to say that "*Nahum*" signifies *consolation*, and "*keik*" a *bosom*, or *heaven*, — plainly intimating that those who settled at this place, and, of course, at others bearing a similar name, were to be received into the bosom of consolation; but whether in a temporal or spiritual sense, the doctor has not clearly indicated.

now Salem, round Cape Ann to the Merrimack River, and up to the head of each of these, across from the head of one to the same point on the other, and including the islands within three miles of the coast. The name of Mariana District was given to the grant. About this time Mason and Gorges determined to unite their fortunes, and jointly obtained a patent of the territory between the mouth of the Merrimack River and the Sagadahoc, extending northward to the great lakes and the St. Lawrence River. This territory was called the Province of Laconia. Having possessed themselves of that large tract of country, these two energetic men set to work in earnest, hoping to realize, by developing its resources, some of that wealth of which they had been dreaming, and which had been the motor in driving them, with all their great energies, headlong into this scheme. Early in 1623 they sent over from London a colony which landed at the mouth of the Piscataqua, and called the place Little Harbor. Proceeding up the river to the site of the present city of Portsmouth, they named the place Strawberry Bank. In the absence of any known reason for applying the name of this delicious berry, historians infer that they must have found the blossom or the fruit of the strawberry exceedingly plentiful.

The colonists divided into two companies, — the first establishing a permanent settlement near the mouth of the Piscataqua; the other, proceeding some miles further up, located at Dover. They were commonly known as the upper and lower plantations, and were in most respects, separate communities, and subsequently lived under distinct governments. Thus it will be seen that New Hampshire was settled in less than three years after the landing of the Pilgrims at Plymouth, in Massachusetts.

Very soon Mason and Gorges divided their possessions, — Mason relinquishing all right, title, and interest in the wild region east of the Piscataqua to Gorges, which took the name of Maine. In consideration of this, Gorges offset and confirmed to Mason all that portion of the Laconia grant west of the Piscataqua, to which he, Mason, having come from the County of Hampshire, in England, gave the name of New Hampshire. So Capt. John Mason, whatever may have been the motive that prompted his action, may be justly regarded as the father of New Hampshire. Under his patronage, and by his influence, efforts, and exertions, the first colony was founded, which, though feeble, was the germ of a sovereign State that numbers a population of three hundred and fifty thousand enlightened people.

Ever since its discovery, America had been a land of romance; the most incredible stories were told and believed; delusion seems to have taken the place of reason in the minds of the people of Europe. The islands, lakes, rivers, mountains, and forests were entirely different from those in the Old World; the very sands were said to be mixed with the precious metal; the islands were represented as elysian fields of perpetual loveliness, and fit abodes for fairies; the rivers, crystal streams whose beds were pebbled with the richest gems: the wonderful natural beauty of the great silver lakes was indescribable, studded with emerald islands, some of them many leagues in extent, abounding in game of the choicest varieties, while the waters were filled with an inexhaustible abundance of the richest fish; and in the great forests grew, in spontaneous profusion, the most valuable wood, while shell-fruit, and many other delicious varieties, loaded the branches, and game of every kind, great and small, was so plentiful that the sportsman's occupation would be gone, as it would kindly approach the habitation of the colonists to be slaughtered at his convenience. This river* was said to be "a faire large river, well replenished with many fruitful islands; the ayr thereof is pure and wholesome; the country pleasant, having some high hills, full of goodly forest, and faire vallies, and plaines fruitful in corn, vines, chestnuts, walnuts, and infinite sorts of other fruits; large rivers well stored with fish, and environed with goodly meadows full of timber-trees." In the great lakes were said to be "faire islands, which are low and full of goodly woods and meadows, having store of game for hunting, as stagges, fallow deer, elkes, roebucks, beavers, and other sorts of beasts, which come from the mainland to the said islands. The rivers which fall into the lakes have in them good store of beavers, of the skins of which beasts, as also of the elkes, the savages make their chiefest traffique; the said islands have been inhabited heretofore by the savages, but are now abandoned by reason of their late wars, one with another. They contain twelve or fifteen leagues in length and are seated commodiously for habitation in the midst of the lake, which abounds

* The Merrimack.

with divers kinds of wholesome fish. From this lake run two rivers southward, which fall into the eastern and southern sea-coast of New England."

It may not be wondered at that the people of Europe should entertain extravagant ideas of the New World, when Gorges himself, a man naturally endowed with a gorgeous fancy, should write such descriptions for the information of the people at home, or that those who sought the New World should do so with a reasonable expectation of obtaining wealth by exploring the land, rather than by the old-fashioned slow and uncertain process of tilling it. Among those who entertained the greatest expectations of this kind was Capt. Mason himself; he knew that the Spanish expeditions to the western hemisphere had returned to Europe heavily freighted with gems and precious metals. The Spaniards had reported that gold was so abundant that the natives, ignorant of its great value, freely used it in the manufacture of the most ordinary ornaments and implements. "Why do you quarrel," said a young cazique to the Spaniards, "about such a trifle as gold? I will conduct you to a region where the meanest utensils are made of it." Mason had learned that gold was obtained from the mountains. New Hampshire was full of mountains; and, reasoning from analogy, bright visions of untold wealth perpetually appeared before him. In his vivid imagination, mines of incomputable riches reposed under the granite base of each great hill, waiting only the magic touch of his masterly energy to arouse it, when it would come forth and invest him with the title, dignity, and prerogative of a feudal lord, while the inhabitants of his extensive domain should merely be his vassals. Such was Laconia, or, at least, the ideal Laconia. The real Laconia was yet to be developed by unwearied exertion, years of toil, patience, endurance, and privation.

After ten years of hardship the disheartened colonists found their condition and prospects no better than at the commencement. This, instead of discouraging the proprietor, seemed to have the effect to render him, if possible, still more sanguine; adversity might darken his prospects, but it only served to strengthen his faith. Chimerical as were his ideas he seems to have been restrained by a most curious mental organization from realizing it, nor does it appear, in all his checkered colonial career, that he secured scarcely a penny's worth of prosperity to a pound of adversity. He died in 1685, with a firm reliance to the last moment, it appears, on the ultimate fruition of his cherished hopes.

In 1629, Passaconaway, a great chief, sold the territory extending from the Piscataqua to the Merrimack River, and from the line of Massachusetts territory thirty miles into the country, to Rev. John Wheelwright and his associates. The deed was signed by Passaconaway, the Sagamon of Pennacook; Runnawit, the Chief of Pawtucket; Wahangnonawit, the Chief of Squamscot; and Rowls, the Chief of Newichewannock, and properly witnessed.

Wheelwright was a very pious and able man, residing at Braintree, in Massachusetts. Endowed with a generous liberality he resolved to overthrow the priestly despotism over the mind and break the strong shackles which bound men to the sway of bigotry. Imbued with this spirit and determination, he presently encountered insurmountable obstacles.

"Forefather's-day" orators (of the spread-eagle genus) — those interested and disinterested, informed, misinformed, and uninformed, sometimes the wise, and oftentimes the otherwise — magnanimously declare that the Puritans, moved by the genuine spirit of religious toleration, fled from homes endeared to them by the associations of a lifetime and sought the wilds of America, there to establish on an enduring basis the great fundamental principle of freedom which they so much loved and had sacrificed so much to maintain, — the principle of religious toleration.

The opinion seems to have obtained that the Puritans, pained and disgusted with the blind spirit of intolerance of religious differences in the Old World, sought the New, where, with the power in their own hands, the corner-stone of the new government should be freedom of conscience, and this immutable principle thus preserved, the Christian millennium should come in America, where the lions of the established church and the lambs of dissenting faith should lie down together, and only the little child of Truth should lead them.

This view of their disposition, judging by a careful and unbiased examination of the records and history of the Puritans, would seem to be entirely erroneous. It really appears that, instead of being deprived of the privilege of conscientious worship, they left the old country rather because they could not persuade or restrain their fellow-subjects to endorse and adopt their own peculiar religious tenets. Thoroughly convinced and honestly believing, without doubt, that theirs was the only true faith, it was sincerely offensive to them that men should repudiate alike themselves and their creed; that, having embraced a heresy, they should wilfully persist in being heretics in spite of the luminous example, and the powerful, untiring, and convincing moral suasion of the dissenters. But so it was, and in casting about for some consistent and efficient means to remedy the moral obliquity of mankind, it would seem they determined to locate where they would have the power to apply such needed and wholesome restraints as the natural perversity and frailty of the human mind required. They resolved to go where they could establish a mitigated or modified form of inquisition, — where, if heretics came, or rose up among them, they could hurl the anathemas of the church; should that not suffice, they could even subject them to such mild legal discipline as whipping, banishing, cutting off ears, and hanging.

All of these punishments and many others have been inflicted and endured for conscience' sake in New England. (O Religion! what cruelties have been inflicted in thy name! How many monsters and how many martyrs have been seen in thy holy cause!)

Soon after the consummation of the Wheelwright purchase a fierce and violent religious controversy broke out, creating a schism among the Puritans of Boston, and leading to an open rupture in the State. They were all dissenters, but the dissenting dissenters were called Antinomians: the latter questioning the policy and disputing the right to force men to the adoption of any particular religious belief by legislative enactment, declaring that, as they were free in conscience, they were bound in religion and in honor to maintain the same right for others, whether inside or out of the close corporation of Puritanism. The Originals, as they were called, in other words, the primitive Puritans, while they declared for religious freedom and toleration, practically controverted their own professions, as the Antinomians declared, by taking the ground and promulgating first the infallibility of their own creed and the right of every man to embrace the truth. Ergo, all who worship with them should and did enjoy perfect religious freedom, while those who did not, deserved and should receive condign punishment. Thus will be seen by their own showing the sum of Puritan toleration.

Meanwhile the contest waxed furious. Wheelwright and his sister, Ann Hutchinson, famous for her great intellectual abilities and persuasive eloquence, were among the foremost leaders of the Antinomians in this memorable theological controversy, and all conditions of people old enough to engage in disputation took sides pro and con in the contest. The gubernatorial election was canvassed with sleepless energy, and with all the acrimony common to zealots. The standard-bearers were Winthrop and Vane, and after one of those heated and embittered contests, unknown except where religion is an element, Winthrop was elected. The legislature immediately set to work to mete out merited punishment to the defeated and dispirited Antinomians. Wheelwright was banished, and went, with some of his followers, to the wilds of New Hampshire, where he settled on his previous Indian purchase at Exeter.

Here this little band of exiles formed a church, and established a government, which may be considered the first genuine democratic form of government ever established in America. The rulers were elected by the people, and the people were sworn to support them; all laws were made by the people in popular assembly, and any proving a public evil were immediately repealed in the same manner. Every man had a voice in the government under which he lived; the source of all power was vested in the people and reserved to them.

" This little association of exiles I consider to be the first institution of government in New Hampshire." *

The sword of persecution lopped off this most thrifty scion from the parent stock, which, transplanted with care and nurtured by the earnest efforts of great and good men, has grown to be a symmetrical and majestic tree, affording a generous shade and protection to the sovereign, intelligent, and progressive people of an independent State.

" Thus the motives of the first settlers of Exeter were in harmony with democratic principles of government. They were exiles ' for conscience' sake.' They came to the wilderness for freedom; they were tried in the school of misfortune; they were disciplined by struggling with persecution. Such was the Exeter settlement. Christianity presided at its birth and ' rocked its cradle.' "† " It grew up; it put forth hands with increasing strength, and displayed in its

* Trumbull. † Bancroft.

form the beauty of youth; it ripened to maturity; it became the State of New Hampshire, — a member of that union which binds together a mighty confederate Republic."*

A union having afterwards been consummated of these settlements with Massachusetts, and the latter having extended her jurisdiction over them, Wheelwright was no longer secure; he fled from the power of his persecutors, and located at Wells, Maine. In the course of time he was graciously permitted to return, which he did, to Hampton, where he resumed his ministry. He afterwards, when Cromwell, who was his personal friend and school-fellow, was in power, went to England and obtained an audience. Cromwell recognized him as an old friend, and was greatly pleased with the interview. Turning to the gentlemen about him the protector remarked, "I remember the time when I have been more afraid of meeting Wheelwright at foot-ball than of meeting any army since in the field."

Wheelwright found himself in favor, and received a gratifying appointment. After the restoration, he returned to New England, and died in 1680, being upwards of eighty years of age.

Sometime about 1660 a frightful and fatal delusion broke out among the New Hampshire and Massachusetts colonists, which was known as witchcraft.

Persons, principally old women, it was believed, colluded with the devil, by which the demoniac qualities and powers of his satanic majesty were imparted to them, enabling them to appear in strange, fantastic shapes to the terror-stricken community, such as flying through the air astride of a broomstick; appearing unexpectedly in unusual places in the form of a black cat or some other animal, also in many unaccountable guises; possessing the power to disappear and reappear at will through key-holes, knot-holes, and other impossible apertures, as well as to afflict any person in the community at pleasure with strange maladies, such as fits and painful contortions of the body, burns, chills, callousness, choking sensations, — in short, a thousand and one maladies that flesh is not properly heir to, and which the most scientific disciples of Galen could neither remedy nor relieve.

As an illustration of the senseless and unreasoning frenzy which pervaded all classes on this subject, one of the witch trials at Portsmouth

* Barstow.

may be given, and it may also serve as a curious relic of an epoch of fanaticism in our history, for the contemplation of a more enlightened posterity. Good-wife Walford was brought before the court, on the accusation of Susannah Trimmings, who was the first witness, and testified as follows: —

"As I was going home on Sunday night, I heard a rustling in the woods, which I supposed to be occasioned by swine, and presently there appeared a woman, whom I apprehended to be old Good-wife Walford; she asked me to lend her a pound of cotton; I told her I had but two pounds in the house, and I would not spare any to my mother; she said I had better have done it, for I was going a great journey, but should never come there.

"She then left me, and I was struck as with a clap of fire on the back, and she vanished towards the water-side, in my apprehension in the shape of a cat. She had on her head a white linen hood tied under her chin, and her waistcoat and petticoat were red, with an old gown, apron, and a black hat upon her head."

Her husband, Oliver Trimmings, was then called and said: "My wife came home in a sad condition; she passed by me with her child in her arms, laid the child on the bed, sat down on the chest, and leaned upon her elbow. Three times I asked her how she did; she could not speak. I took her in my arms and held her up, and repeated the question; she forced breath, and something stopped in her throat as if it would have stopped her breath. I unlaced her clothes, and soon she spoke, and said, 'Lord, have mercy upon me; this wicked woman will kill me;' I asked what woman; she said, 'Good-wife Walford.' I tried to persuade her it was only her weakness; she told me no, and related as above, that her back was a flame of fire, her lower parts were, as it were, without feeling. I pinched her and she felt not. She continued that night, and the day and night following, very ill, and is still bad of her limbs, and complains still daily of it."

Nicholas Rowe made oath as follows: —

"Jane Walford, shortly after she was accused, came to the deponent in bed in the evening, and put her hand upon his breast, so that he could not speak, and was in great pain until the next day. By the light of the fire in the next room it appeared to be Goody

Walford, but she did not speak; she repeated her visit about a week after, and did as before, but said nothing."

Eliza Barton testified that she "saw Susannah Trimmings at the time she was ill, and her face was colored and spotted with several colors; she told the deponent the story, who replied, that it was nothing but fantasy; her eyes looked as if they had been scald."

John Puddington, the next witness, made oath that "three years since Good-wife Walford came to his mother's; she said that her own husband called her an old witch; and when she came to her cattle, her own husband would bid her begone, for she did overlook the cattle, which is as much as to say in our country bewitching."

With respect to the character of the testimony that is given in this case, it is believed to be in sense and pertinence above the average of evidence given in those cases where persons were convicted of the atrocious crime of witchcraft. Although no person suffered the extreme penalty of the law in New Hampshire for it, still this dismal episode in her history would arrest the attention of its student, and furnish a most gloomy repast, only that the same era in the history of a sister State exhibits a catalogue of judicial crimes which totally eclipses the comparatively mild form of this nondescript malady in Portsmouth and neighboring settlements.

In many places, Topsfield, Andover, and especially Salem, in Massachusetts, the virulence of this disease threatened to combine the dire results of those unwelcome triple guests, war, pestilence, and famine. Accusations were hurled against the youthful, the harmless, and the unsuspecting; indictment and arraignment followed close upon their track with the activity of a well-trained hound, and to be accused was synonymous with conviction. As in the case of every popular heresy, reason was deaf, and justice was deaf as well as blind. No cross-examination and no defence were permitted; the bulwarks of justice and protection seemed to have been carried away by the violence of the popular tempest.

Ministers of the gospel precipitated the weight of their character and influence into the seething caldron of unparalleled madness. Judges on the bench permitted and united in the jeers, taunts, epithets, and clamor of the spectators against the defenceless prisoner, instead of restraining them. In some instances, the prisoner, appalled by the terrible character of the charge preferred, knowing the power of evidence which intimated much, though it proved nothing, and feeling assured that certain conviction awaited him, whatever the charge might be, pleaded guilty; and, however strong the evidence and proof of witchcraft against the unfortunate prisoner, contrary to the rule founded on antiquity and common sense of convicting the prisoner on his own free and voluntary admission of guilt in open court, was invariably acquitted. Detectives were employed, whose business it was to ferret out among the obscure, lowly, and unprotected of the community cases of malignant and incipient witchcraft, whose emoluments depended wholly or partially on conviction. That these individuals would not be overscrupulous the experience acquired in the employment of similar agents in the enforcement of obnoxious modern legislation sufficiently demonstrates. There was no such offence known as perjury; malice could do its worst with impunity; no allowance was made for mistaken identity, in seeing, hearing, or memory; the prisoner was considered guilty until proved innocent, and all else failing to convict, the judges often sought to entrap him into damaging admissions.

The community furnished victims, and the officers of the law, instead of seeing that "equal and exact justice" was done, conspired with them in the horrid sacrifice, and in Salem alone upwards of a score of persons suffered execution for offences which reason declares they never had the power to commit. So it went on, without any perceptible abatement of the disease or diminution of the number of its victims, until accusations were made against persons of wealth, standing, and influence; doubtless by some person who had the sagacity to discover that this was the only feasible method of checking the delirium. This had the desired effect, — the spell was broken, the disease abated, and was finally cured.

Such is a faint outline of witchcraft. Ignorance and superstition were probably its corner-stone; mental and nervous affections were undoubtedly superinduced, which inflamed and intensified the malady, removing the bounds to its contagion and the limits to its fatality. Witchcraft may, perhaps, be regarded as an appropriate interlude between the great moral drama of "Puritanism rampant" and the tragic after-piece of desolating Indian wars.

That it would be desirable to cover this period in the history of our country and race with the pall of oblivion, no one will question;

it is true they have the mitigating circumstance of ignorance to plead in extenuation of their folly, and perhaps the future historian may kindly veil their faults and shortcomings with the mantle of charity" when he contrasts the flimsy fanaticism of that dark age with more expensive and fatal displays of this ungovernable element by a people elevated and enlightened by the experience of two hundred years, above the condition of their deluded and misguided ancestry.

Having briefly glanced at the character of the soil, climate, and productions, also the aboriginal inhabitants, as well as the immigrant settlers, and likewise having noticed the discovery of the American continent and of New Hampshire, noting on the record of events the discovery of the Merrimack River, referring only incidentally to subsequent events, it remains but to observe how each successive wave of colonization rolled back farther and farther from the rock-bound coast; how the steadily increasing population swarmed through the forests and over the hill-tops, meeting and overcoming in their resolute advance a frowning wilderness filled with hostile foes, and leaving behind them, in their resistless and triumphant march, a reclaimed and blooming territory, blessed with a government of laws and not of men, nor even yet of mere brute force, until they reached the eastern bank of this noble stream, forming an unbroken cordon of this improved pattern of humanity, extending from its mouth along the Merrimack far up among the mountains towards its source.

When the stalwart and hardy pioneers approached the Merrimack, they found the giant forest-trees still casting their shadows upon its bright waters as they had done for ages past; they found the dusky barbarians flitting like birds of evil omen about the dark and secluded dells and gloomy recesses of the woods, or shooting out noiselessly from obscure pools and eddies upon the broad bosom of this stream in the primitive birch canoe, or vainly endeavoring to excel their rival, the beaver, in skill and dexterity, in the construction of dams, or "stopping-places" for securing fish. They found the Merrimack rolling towards the sea as it had done for all time, and will continue to do forever, — and they found a prize. Here on this river's brink, civilization and barbarism met, — light and darkness, — day and night struggling for the mastery; and who could doubt the result? On the one hand was the rude, untutored child of nature, with no higher aspirations than to follow withersoever she led, in all respects only her simple child; while on the other, appeared her stern and determined master, compelling her to yield of her generous but yet unmeasured bounty a supply for all his wants and comforts.

Civilization found this beautiful river, with its expansive border of timber and fuel trees only useless cumberers of the ground; the broad, alluvial intervals untilled and unproductive except in malaria, noxious weeds, and venomous reptiles; the numberless waterfalls along its course and that of its many tributaries entirely unimproved; the great and now famous places, such as Pawtucket and Namoskeag, a desert waste, — those unsurpassed water-powers being simply Indian eel-pots, — while the pellucid current of the river itself was but a highway or thoroughfare for migratory tribes of fishes, and roving unsettled tribes of savages,—a kind of grand trunk road connecting with the great principal trails at the grand junction, the Winnipesaukee, leading from Maine and the provinces, down through New Hampshire, to the confederate tribes located upon its lower waters in Massachusetts.

They found this beautiful river stretching its magnificent proportions from the convocation of towering, cloud-capped summits on the north, directly through the centre of New Hampshire, dividing the State into two nearly equal parts, pouring over the border and rolling on through a considerable extent of the sister State of Massachusetts: there dropping the generous tribute of its collected waters into the broad bosom of the mighty deep; and they recognized in the quality of transparent and unsurpassed purity which it possessed the elements of health and beauty. In the majestic falls they discerned a mine of wealth richer that Ophir, and as enduring and inexhaustible as time itself. In the liberal volume of its current they saw there was a mighty power, which, under efficient management and control, was sufficient to force away the incubus which hung upon the water and the land, and whirl the wheels of progress and prosperity with busy, pleasant hum.

The first partial survey of the Merrimack River disclosing its inestimable value and importance for manufacturing purposes, it may, perhaps, be said, cast the die. From this time forth the red man made his compulsory bow to the inexorable logic of events, and facing westward with steady advance, — his speed frequently accelerated by the uncomfortable and dangerous proximity of his exterminator, his last remaining hope being to reach the land of sunset, — he left this river and this land, the home of his boyhood and his manhood, his only patrimony, the sacred resting-place of ancestral dust, the pleasant and endearing associations of time, places and events, records and traditions, the old familiar haunts of his people; above all of which, lacerating his obdurate heart and filling his benighted soul with pangs before unfelt, was being forced and torn, as it were, from these places which afforded him his chiefest pleasure and his food, — "The smile of the Great Spirit;" "The beautiful waters of the high place;" "The crooked mountain waters;" and "The beautiful island river with the bright, strong current and pebbly bottom."

But it was useless for him to struggle against the immutable decree of fate, and so he left all of these; the sceptre of his wilderness empire fell from his grasp, his crown crumbled, his ancient power and hereditary rule and supreme kingly prerogative were stripped from him, and he was sent forth a beggar, an outcast, and a vagabond, to be a stranger in a strange land.

Thus departed the aboriginal proprietor, while the march of intellect, enterprise, skill, industry, and progress amply supplied his place. Solitude no longer reigned supreme, or brooded over river, hill, and dale; the vigorous stroke of the woodman's axe resounded through the forest; roads were made; the log-house and the school-house sprang into existence almost together; the little church reared its tapering spire as if pointing out to sinful man the way to heaven, to God; the saw-mill creaked and grated in harsh, unmusical cadence in many localities along the lesser tributaries of the Merrimack; hamlets grew to be villages, and villages towns. Skill, labor, and capital, that all-powerful triumvirate, united their fortune and interest for the mutual benefit of all concerned, and, with industry under intelligent direction for manager, pushed steadily up the river, dispensing wealth on every hand, and building cities, tilling and fertilizing the soil, reclaiming the rich, alluvial intervals, while even the waste places of rocks and swamps and sand were their most fruitful vineyards. Cities, and enterprises involving the employment of millions of capital and multitudes of people, sprung up, as if by magic. Every valley and hill within the radius of this river's salutary influence produced its complement of beautiful and noble women, as well as great, good, and brave men; and this river gave to the manufacturers along its course an opulence of fame for the unequalled variety, quality and value, which is the property and just pride of the nation.

Let us endeavor, then, to realize something of the independence and affluence of the people of New England, to which this river has so generously contributed; observe and note the prosperity and advancement of these two States in intelligence and in all the elements of material greatness, and inquire what portion of it has been contributed by, or is due to, the unparalleled capacity and usefulness of the Merrimack River, and how much of benefit these people may confidently hope and expect to derive steadily and perpetually hereafter from its unstinted, unabated, and inexhaustible munificence.

CHAPTER II.

The Pemigewasset. — Mountains in Summer. — In Winter. — Source of the Merrimack. — Method of taking Salmon and Trout. — Franconia Notch. — Echo Lake. — The Flume. — The "Old Man of the Mountain."

THE Pemigewasset is the name of the principal stream which, at the forks in Franklin, takes the name of the Merrimack. This is a corruption of the Indian word Pemegewasset, signifying "the crooked mountain-pine place," from Pennaquis (crooked), wadchu (a mountain), cooash (pine), and auke (a place). By the adoption of this name the Merrimack was to that extent curtailed and shorn of its fair proportions, for substantially it is the same river in every respect. It seems, then, not only a geographical outrage, but really a deliberate attempt to belittle this lovely river by requiring a large section of it to bear another name; however, as "the rose by any other name would smell as sweet," so the Merrimack, call it what you please, is the same most important stream, the same beautiful and transparent water, the same pride, and honor, and glory of the State and of New England. It stands pre-eminent, and without a parallel the greatest manufacturing river in the civilized world.

Not only does this stream attract the notice and attention of the capitalist, the mechanic, the farmer, and the speculator, but its picturesque and romantic beauty, the wonderful physical phenomena spread out from its pellucid bosom, attract the attention of the man of leisure, the tourist, the student and admirer of nature, and even the true sportsman, who finds in this river and in the numberless tributaries, large and small, and in the fields and the magnificent forests stretching away for many miles, a great abundance and endless variety of fish and game, and, above all else, occasion and cause to offer up a spontaneous tribute of admiration to the great Author of all these wonders and beauties and blessings.

The Merrimack River cannot boast of the gigantic size of the Amazon, the Mississippi, or the St. Lawrence; nor of the white wings of commerce spread upon its bosom, nor is it a vast stream of filthy fluid upon which thousands are transported daily by means of high-pressure engines; but it is believed to be one of the most conspicuous illustrations in the whole range of nature of a grand and generous combination of the useful and ornamental. This river is nearly two hundred and sixty miles long by its course, and is peculiarly symmetrical in its entire extent; its waters are naturally incomparably pure and transparent; and its whole career, from the towering mountains where it takes its rise, to the boundless ocean, is a continual succession of silver cascades, sparkling ripples, broad, calm, mirror-like waters, or romantic, majestic, and useful waterfalls.

The principal value of this river is centred in its grand falls or mill privileges, which in the main stream and very many of its tributaries are quite numerous.

Although it has been said truly that these falls give the Merrimack its principal importance, still it by no means comprehends its complete list of capabilities or usefulness; for, in addition to the many and various other purposes for which it is used, nearly ten million feet of round lumber are annually carried by this convenient highway beyond the limits of the State, in addition to what is moved from the timber section to intermediate places in the State, there to be wrought and fashioned by the buzzing saw and keen-edged plane, aided by busy and cunning hands, into every description of building stock, and useful and ornamental wood-work of every conceivable variety. Thus it may be said that this river presents that rare combination of use and attractiveness along its whole course which renders it, *par excellence*, the most magnificent stream in the world. Other rivers have their grand and remarkable features, famous falls, sublime to look upon, even grand beyond description; but here the record of their qualities terminates. Others receive the homage due to their enormous size, and, though ships may spread their canvas on their surface, still their waters are a sort of filthy mud gruel in appearance and consistence, and are, besides, throughout their whole extent an insatiate graveyard, — a paradise for alligators.

Of the great rivers of the world the St. Lawrence is unquestionably the purest, the most picturesque and romantic, and conveys, perhaps, the largest tribute to the never-sated ocean; but here its excellences end; for, aside from being an outlet for accumulated waters, and a highway for an occasional shipload of immigrants from the Old World, and a conveyance for lumber, it is entirely valueless, and worse. It necessitates an outlay for protection from floods and ice, expensive bridges, and the employment of a large number of that unprofitable, unproductive, and expensive class known as soldiers.

How different it is with the beautiful Merrimack! — rolling along between its natural banks without encroachment, affording sites for cities, factories, and shops, dispensing wealth on every hand, giving employment, and, consequently, health, character, and happiness to many thousands who are employed in the peaceful pursuits which tend to carry a high civilization, comfort, and happiness to the remotest lands; and the high order of intelligence, skill, progress, and prosperity attained by the thriving communities along the course of this stream may be in part attributed to its unparalleled usefulness.

Nor is this its only merit. From the time when the first hardy pioneer set his foot upon this rock-ribbed soil, and set a stern face towards the unbroken wilderness and its inhabitants, this river is commingled with its history as with the ocean itself, and has full many of those daring deeds, and terrible slaughters, and fearful tortures, and hair-breadth escapes, which entitle its contiguous territory most emphatically to the distinctive appellation of the "Dark and Bloody Ground."

As will be seen, history has had much dealing with this river, and it has been celebrated in story and song; and, though much has been said of it in so many ways by so many admiring people, still its incomparable beauty and value are, if not unknown, certainly not appreciated to the fullest extent even yet. It does not appear that the Indians had any regular and permanent settlement at the head-waters of the Merrimack, although it was in the direct thoroughfare between the tribes of Massachusetts, the lower Merrimack, and Canada.

The Pennacooks and confederate tribes traversed this region constantly from Massachusetts to the St. Francis country, often lounging about the head-waters of the Androscoggin and the Connecticut, where it may be said they maintained permanent settlements, and kept in this manner and in this direction a line of retreat open, for such emergencies as frequently arose, down towards the sea.

The upper waters of the Merrimack were, no doubt, a sort of *caché*, or perhaps hotel, where, in the fastnesses of Franconia, in the almost impenetrable seclusion of the mountains and the wilderness, they could rest from the weary march, and from the Pemigewasset and the forest procure such sustenance as they desired.

Having thus obtained a glimpse of the red man's Merrimack, let us see what it is to the pale-face.

This river has its source high up among the mountains, in the very heart of the most romantic country that human eyes ever feasted on. Mountains are thrown together in promiscuous confusion, of all shapes and sizes, — exaggerated haystacks and gigantic pyramids. Some of them remarkably symmetrical and clothed in luxuriant forests, completely enveloped in a dense, rich foliage; others seamed and scarred, displaying unquestioned proof of the ravages of time and the elements; others still, rear their naked and bald summits far up among the clouds, where their furrowed and jagged pinnacles exhibit a poverty of vegetation, not a shrub or a blade of grass to redeem it from total and utter desolation.

In midsummer a tour among these everlasting hills is an indescribable treat to any one; but to the lover of nature, to him who worships in spirit and in truth in the majestic temple of the Almighty, it is a pleasure whose depth and breadth cannot be gauged; for here, spread out before him like a sermon, are the wonderful power, the greatness, and the glory of God. Here is no costly edifice, with a solemn, gray old man to wait on strangers; no tapestried pulpit, with tinsel ornaments and tassels; and no grand organ, with a master to move the keys. Here is the grand proscenium of the throne of the Great Jehovah; the curtain is raised, and he who worships here walks boldly in, unushered and unannounced save by his own affinity. Here the sermon comes to all the senses from the emphatic, comprehensive and persuasive tongue of nature. No salaried clergyman shows off his college polish in rounded periods; but the rocks, the rippling waters, the great forest-trees, all the surroundings, — in short, everything proclaims the eloquent and impressive sermon of God in words of undoubted definition; while the babbling brooks, the rushing torrents, and the gentle gales, sweeping through the wide-spreading branches of the trees, all combine to sing His praise.

What a glorious place of worship is this for those whose limited means exclude them from the fashionable churches!

Here there is no animadversion on dress, or ambiguous reference to taste. The Divine Presence effectually excludes any and all unworthy feelings, preparing the mind properly to commune with the Eternal.

Having stood in the midst of these sublime surroundings, under a genial sun, when the whole face of nature presents its most beautiful and inviting appearance; when the laughing waters sport in the bright sunlight, and the boundless forests are dressed in their full and lovely emerald foliage, and the health-giving and refreshing breezes sweep down from the north, cooling the fevered brain and quieting the overtaxed nerves, — having seen and enjoyed all this, it may not be uninteresting or unprofitable to reverse the picture. Although the view in summer is so indescribably grand, it may be truly said that whoever has not seen the mountains in midwinter will surely fail to realize the awful grandeur of these great realms of solitude.

Approaching the mountain regions, a superabundance of snow is seen on every hand, — the country, as far as the eye can reach, being one vast, unsullied common, the snow being so deep that the top rock on a wayside wall is, like an honest man, a rare sight. Nearing the mountains, the mind is forcibly struck, annoyed, and irritated with the awful silence which now supremely reigns. Arrayed in robes of spotless purity from crown to base, these unstained mounds appear like whited sepulchres; no man nor living thing invades their frigid quiet.

Standing in moody and solemn grandeur, these towering cones glitter in the pure sunlight, and seem, by some indefinable fascination, to allure unthoughtful mortals to their doom. Were the 'prince of the powers of darkness enthroned on these inhospitable and glittering pinnacles, his empire could not be more secure from intrusion than is that of the king of frost, aided by his most efficient ally, old Boreas, whose cheeks, it would seem, must ultimately crack by reason of continuous distension. Silence and desolation reign supreme throughout these empires of solitude; and it would seem as if the Almighty had here erected an impassable barrier to the contaminating contact of man with this bright and beautiful land of gloom. The fearful stillness of death broods over all. The very view of this glittering but ghastly territory is frightful, and man avoids it as he would the deadly shade of the poisonous Upas. Even the wild beasts of the forests creep stealthily around the frozen base of this polar Malakoff, not caring or daring to risk a conflict with a foe so formidable and invincible; and the hardiest wild bird shrinks away down into the thickest and best protected forest, or plunges boldly under the snow to escape the marrow-chilling breath of the dread monarch of this land of terror and loneliness.

High over all, at times, the cold, gray sky looks down without a sign of relenting; while the atmosphere, misty with frosty distillations, or clear and rarefied, penetrates to the very vitals, biting like a beaver and cutting like a Damascus blade. At other times, the wild, angry clouds roll up in dense, dark masses, enveloping the glistening summits in their angry folds, while the winds howl fiercely and roar savagely through the deep and yawning chasms, as the hurricane tears along the boundless waste of ocean. Then comes the snow, fine and thick, sweeping in great waves along the mountain-side, even far out upon the plain, penetrating every nook and crevice, whirling high in air, then rolling down in thick masses resembling a mighty avalanche, then, turning back, apparently assumes a host of strange, fantastic shapes, reclimbs the rugged, dizzy height, there again to be taken in the arms of a still more gigantic *blow*, and hurled with lightning speed far away, and scattered over the land. Thus is another spread placed on these sleeping giants. The winds abate to a dismal moan, then sadly sigh through the wide-spreading branches of leafless forests, and die away to silence, while night and the cold, round, silver moon together roll upon the scene, intensifying its frightful polar horror.

The Merrimack River has its extreme northern source in the Willey Mountain, — so called in honor of the Willey family, which was destroyed by the fearful slide that occurred on its eastern slope in 1826.

This took place during the night. Hundreds of acres of the almost perpendicular mountain-side, which had been loosened and undermined by long and heavy rains, came pouring down upon the interval, bearing with it huge rocks and the entire forest which grew upon and covered it, burying in its overwhelming rush every member of this unfortunate family, — father, mother, five children, and two servants, all finding a common grave in the *débris* of this terribly fatal slide. It is in this famous mountain that the true source of the Merrimack is located. On the western slope of Mount Willey, directly opposite the slide and near the base, is a pond called Ethan Crawford's Pond, in honor of the old pioneer of the Notch. This pond, located in an almost unexplored wilderness, but little known, is generally believed to be the source of the Merrimack. This, however, is a mistake, for up the mountain, to the north-east of Ethan's Pond, at a considerable distance, and near the summit of Mount Willey, is another pond, nearly the same size of the latter, which is the *true* source of the Merrimack. The waters of this pond escape in a southerly direction, and, swinging to the west, after travelling several miles, unite with the outlet of the Ethan Pond, to the west of the latter, and form the nucleus or starting-point of this magnificent stream. Nor does it long remain the insignificant rivulet we find it here; for, surrounded with lofty mountain ranges on either side, it receives accessions at almost every step, and a progress of but a few miles renders it a real river, respectable in size, magnificent in its surroundings, lovely and romantic beyond description.

Starting from the base of Mount Willey, this stream runs through an unbroken wilderness for nearly forty miles. A country so wild and grand few have ever witnessed. It is a singular if not important fact, that within five miles of the source of this stream may be found the sources of two others, each in its course following a cardinal point of the compass, namely, the Merrimack, running south; the Saco, east, and the wild Amonoosuck, west; while, but a few miles away to the northward, is the Magalloway, otherwise known as the Androscoggin.

Following down the Merrimack, or, as it is commonly called, the east branch of the Pemigewasset, it will be found that a grand spur of the White Hills shuts down within a few hundred yards of the course of the stream, on its right bank; while, on the left, a similar range closes in, leaving a broad and beautiful interval between.

These two gigantic enclosures follow the river, at a somewhat irregular distance therefrom; and, as the stream starts on a due west course, it maintains a regular trend to the southward, so, also, do these mountains, until the Franconia range is met, which continues in the same direction until the foot of the Franconia Notch is reached. Thus, it will be seen that, throughout its wildest course, it is guarded on either side by a huge wall of mountain ranges.

Passing down the stream some six miles from the confluence of the waters of the two ponds, a large brook flows in from the right, and is remarkable for its coldness, purity, and transparency, and also for the immense number of trout which crowd its waters and seem to regard it as a sort of fishy paradise. Leaving this brook, — which is something of a sacrifice, — and reaching a point some ten miles farther down, we arrive at the Grand Falls, beside which the Amoskeag and the Pawtucket, except in volume of water, dwindle into insignificance. Just below is another considerable tributary, and still farther down the Hancock River adds its quota to the swelling flood. All of these streams, large and small, abound in trout and no other kind of fish. Previous to the erection of the Lawrence dam, when every variety of migratory fish came up the Merrimack, the salmon alone, of all the tribes of the sea, sought the cool, pure waters, and the shaded and silent retreat at the head-waters of this river. While on this subject, it may not be uninteresting to observe the nature and habits of the salmon and the salmon trout.

These two noble fish appear to be of the same genus, having substantially the same habits and the same general characteristics. Both are migratory; both present the same general external appearance, being beautifully spotted; and, like the lily of the valley, "Solomon in all his glory was not arrayed like one of these." Both are the richest, finest, most delicate-flavored, and the most athletic, size considered, of all fish.

After the spawning season the salmon goes out to sea, while the trout remain in the rivers and other deep fresh waters during the winter; but there is one remarkable fact concerning the habits of these fish, which is established beyond controversy, — each fish invariably follows up the same stream to spawn where itself was incubated, unless some insurmountable difficulty is placed across its way, such, for instance, as the Lawrence dam.

The method of taking these magnificent prizes was at one time well known, — nowhere better than along the Merrimack. When the fish disappeared the practice was of course suspended; but now that public-spirited legislators and gentlemen have taken hold in earnest to remedy this great loss, and fishways have been constructed to help them pass the dams, and salmon-breeding is come to be a recreation and a profit, it seems likely that our cold mountain streams may be restocked with this admirable fish, and it may not be inappropriate to briefly refer to the various methods of taking them. How the Indians took them it is not worth while to inquire or explain; therefore we will see how expert and skilful fishermen of the pale-faces secure them.

There seem to be but three principal methods of taking the salmon, namely, the fly, the net, and the spear. The spear is tolerably effective, but it takes him at a disadvantage, and also lacerates his flesh, and is in no sense the pastime for a true sportsman. The net is also used by those who make it a vocation; it is circular, nearly two yards in length, and about the size of a common bushel-basket at the opening, which is round, and woven upon a hoop prepared for this purpose; a handle the length of the net is attached by means of the hoop.

Any and all of these arrangements may be varied to suit the necessities of the operator, who proceeds to the falls through which the salmon have to pass, and in the narrow channels between the rocks where the waters rush and foam and plunge, here is placed the net, and soon the powerful beauty rushes in, and, of course, is easily secured, being enveloped in a labyrinth of strong meshes.

For the man who has no object but to secure the salmon for his market value, the spear and the net may answer all purposes, but for the genuine sportsman, who would rather buy the fish, or even do without it altogether, than engage in this tame pastime, the excitement of the fly is the only really satisfactory method. Not only this, but he acquits himself of any accusations of cruelty which his tender conscience may prefer by claiming to display a sort of chivalry, founded on the plea that the fish was entirely free to take the fly or not, as he pleased, and that if he chooses he may test its quality, or inquire what it is made of and take the chances, or if you please the consequences, — declaring that thus they only manifest the disposition of beings claiming to be of a much higher order, and endowed with reason to teach them better, but who, nevertheless, often get themselves into difficulties as inextricable as taking the fly.

In catching the salmon it is very necessary to have the experience and skill of a practical fly-fisher; that is to say, with a rod and line of sufficient length to be able to cast the fly into a six-inch ring at a distance of twelve or fifteen yards, then take position in the bow of the boat, while at the stern sits your skuller. Arrived on the salmon ground the fly is carefully cast, and the skuller is on the alert to back water, or handle the boat in any manner he may be directed. As the fly gently touches the surface, it is taken, if at all, and then is the beginning of a most exciting scene. Of course he is hooked at once, and, discovering this, he gives an exhibition of his unequalled agility and strength. Madly rushing up stream and down — it being important to keep a taut line on him — he is reeled out and in as the exigency requires. Now is the real period of enjoyment for the true sportsman. The excitement of having on his line so rich a prize, the science required to handle the job properly, the danger of his breaking away, the plunging and floundering, the fantastic gyrations, and the intense desire to land him safely, — all conspire to render this sport attractive and exciting in the extreme. It is always necessary to allow the salmon to become exhausted before any attempt is made to secure him, as any effort to complete the capture while he is in full possession of his unexampled muscular power would undoubtedly prove disastrous to the success of the party, and a hard struggle with a healthy thirty-pound salmon might possibly involve more serious consequences.

As has been said, the trout is, to all intents, the salmon modified. There are several varieties of trout, each distinguishable by shape, color, or quality; either of these, and sometimes all combined, are used to recognize one breed of trout from another. These variations from the high excellence of the salmon trout are by many attributed to the water, or, more properly, to the country they inhabit. Thus we have in all the high-land tributaries of the Merrimack, and, in fact, all running water, except the inlets and outlets of ponds, the genuine salmon trout in the highest state of perfection, and even a superlative silver salmon trout; then there is in swampy streams and muddy ponds the mackerel trout, a blue and less rich flavored kind; there is also a variety of trout, very rare, and never found only in small, clear ponds, located at a high altitude, some two thousand feet and upward above the level of the sea. This trout, or at least those examined, vary from one to two and a half pounds, and are peculiarly rich,—being declared by epicures to excel in all respects even the salmon itself. Proper lake trout differ but little in quality, and but for the esteemed name they bear would attract no more, perhaps less, notice than the salt-water stand-by which has given its name to the sandy desert beyond the Plymouth Rock,— renowned Cape Cod.

The universal method of taking the trout seems to have settled down to the rod and line, and if an occasional vandalism is perpetrated, such as liming a brook, or other kindred meanness, it is only an exceptional case, right-minded people frowning down all such doings, and approving the genuine disciple of Isaak Walton.

In pond-fishing, and in other open places, the fly is used at certain seasons, as in salmon-fishing, with good success; in fact, it is sometimes indispensable, and in other localities the bait is caused to *ricochet* along the surface precisely as for pickerel, this being the only means of securing a "bite." These, however, are exceptional cases; the great bulk of trout-fishing is the well-known system of baiting the hook and dropping it in the water.

For bait, the universal India-rubber, mud or angle worm is first and foremost; the yellow-headed white grub, or potato worm, the grasshopper, and several other insects, and sometimes fishes' eyes are used with good success. Many people avoid making a noise when trouting, such as calling to each other, etc. It may be observed in this connection that it is believed trout hear no outside sound, either from a physical disability, or from being in the water. Whether this is or not a correct observation, experience has proved that no mere sound alarms them in the least; but they are extremely sensitive and keen-sighted; a shadow thrown across the brook alarms them at once, or a step on the bank heavy enough to jar and agitate the water affects them instantly, and away they dart almost with the rapidity of lightning. The reason why better luck is supposed to be assured on a rainy day is undoubtedly because the duller the day the less liable is the fisherman to expose himself, and thereby alarm the trout, and, further, the rain beats down flies, and washes from the bank worms and insects. Instinct teaching them that rainy days are their thanksgiving days, they are on the alert, and as ready to take the fisherman's bait, providing he conforms sufficiently to nature to prevent detection, as anything else. As soon as the snow-water is cleared from the brooks, and migration to the spawning grounds becomes general and rapid, the trouting season may be said to have properly commenced; but the best period for a guaranty of good success is regarded as having arrived when the apple-trees are in full blossom, especially when favored by the supposed additional advantage credited by tradition or experience to the influence of the sign Virgo. The only sensible way to follow a trout stream is to follow it down, not up. It is as natural for a trout to head up stream as for mankind to head towards progression, — it is an invariable rule. Being prepared, then, in all particulars, for the sport, seek the stream; if it is frequently fished, great caution must be exercised, and care must be taken to keep out of range of keen eyes, and also not to step heavily on the bank which agitates the water, when the extreme sensitiveness of the trout takes alarm and away he scuds, or secretes himself securely behind some friendly shelter, and nothing can tempt him abroad until he is satisfied the danger is past. If, on the other hand, the stream is comparatively free from the visitations of fishermen, step boldly into the water, allowing the hook to play among the ripples below. Armed with approved bait, a straight, slender, pointed bamboo rod and a basket on the side, one is the prince of fishermen securing the king of all fish, and enjoying an undisputed nobility of sport.

TROUT AND SALMON FISHING.

"Now, when the first foul torrent of the brooks,
Swelled with the vernal rains, is ebbed away,
And, whitening, down their mossy-tinctured stream
Descends the billowy foam, — now is the time,
While yet the dark-brown water aids the guile,
To tempt the trout. The well-dissembled fly,
The rod fine-tapering, with elastic spring,
Snatched from the hoary steed the floating line,
And all thy slender wat'ry stores prepare;
But let not on thy hook the tortured worm,
Convulsive twist in agonizing folds,
Which, by rapacious hunger swallowed deep,
Gives, as you tear it from the bleeding breast
Of the weak, helpless, uncomplaining wretch,
Harsh pain and horror to the tender hand.

When, with his lively ray, the potent sun
Has pierced the streams and roused the finny race,
Then, issuing cheerful, to thy sport repair;
Chief should the western breezes curling play,
And light o'er ether bear the shadowy clouds.
High, to their fount, this day amid the hills
And woodlands, warbling round, trace up the brooks;
The next, pursue their rocky-channelled maze
Down to the river, in whose ample wave
Their little Naiads love to sport at large.
Just in the dubious point, where, with the pool,
Is mixed the trembling stream, or where it boils
Around the stone, or, from the hollowed bank
Reverted, plays in undulating flow,
There throw, nice-judging, the delusive fly,
And, as you lead it round in artful curve,
With eye attentive, mark the springing game.
Straight, as above the surface of the flood
They wanton rise, or, urged by hunger, leap,
Then fix, with gentle twitch, the barbed hook;
Some lightly tossing to the grassy bank,
And to the shelving shore slow-dragging some,
With various hand proportioned to their force.
If yet too young, and easily deceived,
A worthless prey scarce bends your pliant rod,
Him, piteous of his youth, and the short space
He has enjoyed the vital light of heaven,
Soft disengage, and back into the stream
The speckled captive throw; but should you lure
From his dark haunt, beneath the tangled roots
Of pendent trees, the monarch of the brook,
Behooves you then to ply your finest art.
Long time he, following cautious, scans the fly,
And oft attempts to seize it, but as oft
The dimpled water speaks his jealous fear;
At last, while haply o'er the shaded sun
Passes a cloud, he, desperate, takes the death,
With sullen plunge; at once he darts along,
Deep-struck, and runs out all the lengthened line,
When seeks the farthest ooze, the sheltering weed,
The caverned bank, his old secure abode,
And flies aloft, and flounces round the pool
Indignant of the guile. With yielding hand
That feels him still, yet to his furious course
Gives way, you, now retiring, following now
Across the stream, exhaust his idle rage;
Till, floating broad upon his breathless side,
And to his fate abandoned, to the shore
You gayly drag your unresisting prey."

THOMSON'S SEASONS.

The source of the Merrimack River is more than six thousand feet above the level of the sea, and thus it is not strange that it has the quality of purity unexcelled, and its course, through its first forty or fifty miles, is magnificently wild and grand; the silvery, transparent river, meandering gently through the sylvan forest shade, or dashing and roaring through the narrow chasm, or leaping and tumbling in a milky volume over the rocky falls, and then calming down like a mild and still May morning, and settling into dark and quiet pools, where the noble salmon was wont to lay, like a tiger in his lair, crouching, with a keen eye and invincible prowess for the unwary prey. Here, the great blue heron, tall as a well-grown youth, stalks majestically up and down, with an eager eye on the unapproachable trout, more difficult to obtain than the simple perch and shiner among the reeds and rushes on the marshes of his usual fishing-ground; the sleek, sable mink breeds in abundance and security; the great, clumsy, brown bear, though unexpert, tries his skill at diving, trout being for him a rare and savory delicacy; the fish-hawk and the fish-eagle perch above the pure, transparent tide, and, with resistless swoop, bring up the spotted captives. The wild deer comes down from the alpine surroundings, not timid here, and cools and refreshes itself in the pellucid current. Numerous aquatic fowl here also congregate; the little wood duck, the most singular specimen of ornithology, a combination of water-fowl and land-fowl, at home equally on the water, or perched on a forest tree. Silence and solitude brood over all this lovely land, save when the wild beast calls to his mate, or howls fiercely and defiantly along some almost inaccessible precipice, or some bird shoots screaming across the magnificent valley. Huge birds build their nests among the mountain fastnesses, and flap their expansive wings lazily from crag to crag, or rest on a trusty branch of some gigantic forest tree. Conspicuous here is found primeval nature, and all her creatures here unerringly proclaim that man, the foremost of them all, as yet has not appeared, except by proxy; that his cunning and rapacity, his cupidity and cruelty are, happily, unknown in this secluded spot. The mountains here, that range along the stream on either side, are covered with a dense, unbroken wilderness, high up, and it is curious to observe the line of demarcation between the deciduous and the evergreen, regular as the waist-belt of a charming belle,— it marks the boundary of another climate, soil, and production, while the seared, seamed, and scarred summits most clearly show that neither soil nor vegetation there exist. The forests embrace almost every variety of timber. The hemlock is an especial favorite, being browsed by the deer and peeled by the "fretful porcupine."

Tall and stately spruces, straight as an arrow, symmetrical, sound, and plentiful, await the woodman's axe. Game, especially of the larger and less common varieties, abounds in this great forest. Why should it not? Stretching far away to the north and east, it is almost an unbroken wilderness to Maine and Canada. But a few years since the sneaking wolf here prowled for prey. The moose may still be found, and deer are plentiful. Bears are so common as to be often troublesome to the sheepfolds beyond the borders of the wood. Lynxes and yellow-cats are often seen, and not unfrequently captured, and mountain sable are trapped extensively. The mountain partridge is found in great abundance, and of a superior quality and flavor, and there are seasons when the wild pigeons crowd the woods, until the whole surrounding forests are literally alive and swarming with them. There is probably no region of country so extensive, so wild and diversified as this, in any of the older States, certainly not in New England. It commences near Conway, and runs north-easterly to the State of Maine, and north-west to Franconia, bearing north-east again, until, reaching the great forest that skirts the western base of the White Hills, it stretches away to the north indefinitely. This territory is called ungranted land, as no settlements have been made, and no town-lines traverse this grand wilderness of wood, water, and gigantic granite walls. But of all this vast extent of sublimity and grandeur, no section exceeds what may be termed the Fairy Grotto, or the great wild valley of the head-waters of the Merrimack. As an obscure and secluded retreat how grateful, serene, and quiet; how much coveted by the man of business in the great marts, whose brain whirls, and is almost agonized by the tramp of many feet, and the ever-rushing human torrent of countless thousands; for the student, whether of nature or books, who will here receive inspiration, enlarging mind and thought to a more comprehensive, possibly a more common-sense view of God, and, consequently, of man, and his relations and duties to both!

The mechanic and the laboring man may here resort, and, stretched on beds of savory hemlock, surrounded by an atmosphere, health-giving, and pure as the breath of the Deity, relax from toil, refresh and recuperate.

No one can travel hither without receiving benefit. The tourist, the pleasure-seeker, the sportsman, the artist, even the Christian, or he that professes to be a Christian, can here learn the ring of the genuine metal, and, possibly, by contrast, learn the true measure of the scrimped and contracted pattern which he boasts, and better still adopt the other fashion. Here the misanthrope may repair, and if his malady has come to be confirmed, the dismal character of his surroundings may nourish his disease, and thus afford even him a dubious consolation; or, if his disease should be of the mitigated melancholy type, the bright and beautiful around him, fresh and glorious from the hand of the Great First Cause, may wean him from his unseemly, unnatural, and unprofitable despondency, and bring him to a more just and correct view of his duties and responsibilities as a rational and intelligent human being.

Approaching the splendid falls before referred to, one cannot fail to be struck with their great power and wonderful natural beauty. Except in the quantity of water discharged over them, these falls surpass all others on the river, and standing here and gazing on this great power now wasted, one may fancy he sees a prospective Lowell, or an embryotic Manchester, with its monster mills, its rattling machinery, its radiating streets, and its busy people. Although this may long remain a wilderness, unknown to any but daring explorers, it is certain that there is a splendid water-power, and a cheap, easy, and convenient means at hand to secure and control it. The purity of this region is so proverbial, that health, strength, and beauty would conspicuously mark such a community; in fact, it may be supposed, only by waiting for a fatal accident, could that peaceful neighborhood, called a graveyard, be commenced, and its narrow dwelling sites be taken up by actual settlers.

Most people are undoubtedly familiar with many of the physical phenomena scattered about all this region: but as there are some interesting localities which it is impossible for but few to see personally, it may not be uninteresting to mention them. Nearly on the summit of the left-bank range there is a small pond, containing not more than five acres, stocked liberally with the finest variety of trout of large average size; the specimens taken ranging from one to two and a half pounds. This pond is fed by a very generous and pure spring, located about one-fourth of a mile nearer the top of the range, and is discharged over a smooth rock, very steep, for more than six hundred feet, the bed of the stream being from six to eight feet wide, the water, which is of an average depth of one half inch, running so smoothly that it is difficult to discover that there is any water. This is, of course, a fine trout brook; and finds its way to the river some miles from the foot of Lafayette mountain. This pond is east of Mt. Lafayette; south of it there is an extensive windfall, — more than five thousand acres of forest, all spruce, have been blown down, not a tree remaining, and every one taken up at the root, peeling the soil entirely from the bed rock. A not very abrupt ascent of half a mile east brings one to the summit, where a remarkable view presents itself. The eastern face of the range is apparently perpendicular more than three thousand feet, while at that distance below your feet, the country appears perfectly level as far as the eye can reach, — a dozen or fifteen miles. This level country spreads to the north, east, and south, and is covered with beech, maple, and other leaf-shedding trees, without a single opening in this vast expanse, except where the granite crops out above the surface. Through these breaks here and there are seen glimpses of the sparkling waters of secluded lakes, and the silvery cord of the meandering brook, while many miles away to the south-east stands out against the horizon the grim and barren peak of old Chocorua. Westward, seemingly so near you fancy you could hurl a stone upon it, but in fact more than ten miles away, is the naked and frowning Lafayette; to the north-west, the great Haystack, lofty and superb, stands the central figure of a number of gross and ungainly knobs, while over all, some forty miles away to the north, is seen the conspicuous crown of Washington. Camping on one of these great elevations gives one an idea how much earlier " the jocund day stands tiptoe on the misty mountain-top " than in the valley below. Standing here in the early dawn, look down upon the silent valley robed in emphatic night, watch the impenetrable gloom as it fades into an intangible shade, and then to radiant day. As the earliest scintillations of rosy dawn touch the highest summits, and the first beams of the morning sun gild and illuminate first of all creation the highest pinnacles, so does the bright sun of knowledge and of truth first illumine the highest intellectual summits, and gradually but surely penetrate the lower level of the valley of mediocrity, dispelling in like manner the mental and moral darkness.

Franconia Notch, to the east of which, through the wild valley of the Upper Merrimack flow the cool, bright, sparkling waters of the principal branch of that stream, is on the west of this valley, with the Lafayette or Franconia range between the river and the Notch road, running parallel the whole distance from Plymouth to the top or opening of the Notch. This wonderful phenomena of nature is talked about so much by tourists, journalists, and others, is visited by so many thousands, that were it not an extraordinary subject it would have become long since exhausted; but the fact that everything connected with it is listened to with marked attention and eagerly sought for, proves that the subject is still fresh, and that the unparalleled sublimity, grandeur, and romance of this great warehouse of nature's wonders have attracted the liveliest interest among all classes of people. No one sees this gigantic collection of impressive curiosities without a striking and deep-seated impression of their peculiar grandeur. Men may see a great ship, an immense building, or any other interesting work of art, and view it with a kind of admiration; still its salient features are skill, labor, and capital; but here he sees the handiwork, nay, the visible presence of the ever-living God. Here is his great temple, here His throne; here is open to the poorest vision and the dullest comprehension, the majesty and perfection of His works and laws. Many journey to these scenes year after year, attracted by the indefinable emotions awakened in their minds at each successive return.

" I had seen pictures only of this mountain scenery before. Pictures? Mockeries! the best that artist ever sketched but as faintly portrays their grandeur as dew-drops describe an ocean! One exclamation of wonder hardly dies in reverberation ere the eye falls on some new pleasure quite as delightful to behold. One almost fancies that invisible spirits sit enthroned upon those giant cliffs, and are ever preparing a gorgeous dioramic display for mortals' contemplation! His must be a wretched philosophy and worse religion, which fails to recognize the perfecting agency of the great Jehovah in this grand master-piece of natural architecture!" *

Placed by itself, without this remarkable combination of grand surroundings, the Echo Lake, the Flume, the Old Man of the Mountain, or any of the great features of this locality, would stand out in bolder relief than now, and consequently appear to the beholder more awfully grand than as it does, surrounded by so many astonishing and bewildering sights, notwithstanding they are each separately a wonder, and an insoluble problem, and as a group, a marvel, unequalled by any collection of natural objects in the world.

Nothing can produce a more happy or beneficent effect upon the mind than a contemplation of the works of nature. No mind so besotted or so fragile but it can draw inspiration to higher objects, aims, and ends by approaching this pure fountain.

Standing here under heaven's great azure dome, if there be any who know nothing of Him they may at least observe His works, which are on the grandest scale, spread out with a benevolent and lavish hand, and in the Divine Presence at least enjoy a more soul-inspiring, healthful, and innocent recreation.

Come, then, ye frequenters of gilded and gas-illumined palaces, who transpose the order of nature by turning night into day at the expense of health and peace of mind, leave the cup of body and soul destroying beverage; leave your pasteboard and human knaves, chips, shells, spotted cubes, and miasmatic atmosphere, and hasten to the pure fountains of sweet waters and bracing air fresh from the mountain-tops and the fragrant forests, and here gather a loftier ambition, and a more profitable and sensible view of the perfect original grand design!

One of the greatest objects of interest is the Echo Lake, located on the east side of the road near the northern entrance to the Notch. This lake is deep, pure, and tranquil, and is a curiosity and a wonder. Surrounded by lofty mountains and towering cliffs, the report of fire-arms, the speaking-trumpet, or the voice reverberates from crag to cliff in oft-repeated echoes, point after point repeating the sound, returns more and still more indistinct until it has recoiled from the entire circumference. Here people linger for hours listening to this remarkable echo, tracing the gradual diminution of sound; if,

* Extract from letter of D. W. Smith, 1867.

for instance, a question is asked, it will be distinctly repeated usually several times; it then comes back a pleasant murmur, then an indistinct and solemn muttering, murmurs softly, and whispers a final farewell, and is lost in infinitesimal and expanding circles!

Nestled down at the very base of these gigantic surrounding Alps. as if seeking and enjoying a calm repose, hemmed in and fringed with a deep foliage, this lake mirrors the sky and the fleecy clouds, the mountains and the forest, in its tranquil bosom. Its waters are supplied with trout, and skiffs with fishing and pleasure parties skim its placid surface. All sorts of appliances, cannon, speaking-trumpets, etc., are furnished with which to procure every variety and volume of echo, and a day's entertainment may be had here, free, healthful, and interesting, if not profitable. Cannon Mountain has a huge rock poised upon its summit, of such peculiar shape as to be a very accurate likeness of an immense cannon. This monster gun appears to be in position, pointing its black mouth directly across the Notch road, as if prepared to belch forth flame and iron hail upon any who should attempt a hostile invasion of this modern Thermopylæ. Perhaps it is a simile of the "Union Gun,"—a standing and enduring menace to any who would trifle with that sacred compact.

The Flume is one of those works or freaks of nature that is viewed with deepening amazement. The more it is seen and examined the more intensified becomes this emotion, and the question at once arises, How came it so? Conjecture, speculation, and theory are resorted to by each spectator as multifarious as the visitors; luminous but unsatisfactory, for the great problem is still destitute of a definite solution. Was it accidentally left so when America was upheaved from the great world of watery waste in which it was submerged? Was it a deep cicatrice on the plastic face of nature, torn by some rough monster missile, hurled forward and impelled with impetuous and resistless force from the overhanging summit? Was the granite base of the everlasting hills riven by some convulsive throe or throb of the great heart which swells and pulsates in the bosom of mother Nature? No one can tell; neither the man who understands all science and all learning, nor the most inquisitive Yankee can trace the cause; the effect alone is visible.

Nearly three quarters of a mile the solid rock is cleft and rent asunder some twenty feet, the sides being perpendicular and as smooth as the work of a machine, and about fifty feet deep. A bright and noisy brook babbles along the pebbly bottom. Situated in a romantic forest, spanned by several rustic bridges with seats ranged on either side, and as it is completely surrounded and protected by a grateful shade, and a shivery draft of air constantly circulates through its length, it affords a most admirable and desirable retreat for sweltering and suffering mortals when Cancer, blazing with Promethean flame, denotes the torrid solstice.

Notwithstanding the magnitude and sublimity of this remarkable feature, as if upon inspection she was not quite satisfied with her work, nature has prepared a compound wonder. From the summit of the mountain which frowns dubiously upon the Flume, a huge granite boulder has become detached, and, overcoming every obstacle that opopposed its progress, has rolled down and precipitated itself over the brink of this chasm, and hangs suspended near the top by the points of its greatest diameter, so that one passing under it is forcibly reminded of the sword of Damocles, and involuntarily dodges. However, it is not an improvised pile-driver, and there is no danger of being flattened into a broad sheet of gold leaf by its sudden and unexpected descent.

How long it may have been poised and securely held in these powerful clamps is mere conjecture. Perhaps the frosts and snows of many centuries have fallen, lingered, and dissolved upon it; certain it is that nothing remains to mark its pathway hither. Huge trees, that would have been overborne by its resistless weight and force had they been standing, guard every avenue of approach to its present resting-place, and all traces of its track are entirely obliterated. Perhaps it was hurled through the realms of space by some great Ajax of the Western World, which, dropping here, has left no trace, or mark, or sign of its career!

The Old Man of the Mountain is another figure in this group of wonders. This gigantic profile is said to be sixty feet from the forehead to the chin, and creates a powerful impression on the mind of every beholder. It is located at the southern termination of this range at its highest point, and stands out in bold relief, with features as strongly marked and well defined as though fresh from the hands of the sculptor. A front view of the locality simply demonstrates the fact that the mountain shows nothing peculiar or noticeable, only a precipitous and jagged bluff, with furrows and angles; but a side view gives it shape and develops an immense and massive profile, remarkably natural in outline, perfect in feature, and surprisingly symmetrical in the detail and in the total of its entire arrangement. The altitude of this profile is such that its enormous proportions are requisite to give it life-size. Thus, in likeness, workmanship, and perspective, it combines the skill of the master sculptor with the perfection of his art.

The throat is delicately and accurately chiselled, the neck is perfect, the shoulders broad and heavy, the chest deep and massive. This is the Monarch of the Mountains. There he has reigned undisturbed and unquestioned for long, long ages. The frosts of countless centuries have rested and dissolved on his massive brow; still there he sits serene and grim in his weather-beaten granite, and there securely enthroned he will remain through numberless ages yet to come, unless some terrible convulsion of nature shall shake the everlasting hills to their very centre and overthrow his adamantine throne! As it is with the Flume, the impression on first beholding this grand spectacle is striking and impressive, and this impression is deepened and intensified by a closer observation and more thorough study, until at last, turning away with an indefinable awe, the wonder is if nature, with her boundless resources, has ever excelled this in any of her achievements.

Directly at the base of this mountain is a beautifully transparent lake, which is known as Profile Lake, or the Old Man's Wash-bowl. Like all the other waters of this region it is supplied with trout, and is, on this account, and to see the Profile, a great resort for sojourners in this region. The Profile has been so often inaccurately sketched by pen and brush, — "the mirror," as it were, "held so poorly up to nature," — that it would seem impossible to do the subject complete justice. It can only be truly appreciated by viewing its gigantic and symmetrical proportions.

How great are all these wonders! In the pinching frost, in the scorching sun, or in the grateful shade of old primeval trees, let no vain, ungodly brawler say it was all accident and chance, for, by the merest unsubstantial shadows of these majestic peaks, he stands condemned. The foaming cascade, leaping with headlong force from steep to steep in sparkling, laughing, joyous, unrestrained freedom, proclaims his folly, his simplicity, or his incomprehensible depravity!

He who has yet to learn that this is a conspicuous part of the great original design has a lesson simple and easy, which, simple as it is, may overtax and overwhelm his benighted faculties.

Perhaps the most frightfully grand and terrific of all the elemental visitations is a thunder-storm among the mountains. Fleecy vapors, borne on lively zephyrs, scud across the dome; the sun, rolling high, pours a flood of scalding heat, and men and animals are fain to seek the shade. Soon the atmosphere becomes thick, hazy, and oppressive, and the sun looks dull and jaundiced: low, distant muttering, long-continued and oft-repeated, is heard; the sun now appears in a malignant fever; great thunder-heads, — clouds heaped on clouds, — dark and dun, roll up from behind the mountains and shroud their peaks in awful gloom.

Thunder, heaven's great siege artillery, now belches forth in one tumultuous crash, the winged lightnings leap from the angry clouds, darting in fiery and eccentric chains, gyrating about the murky mountain sides and dashing remorselessly upon the devoted head of some old forest patriarch, shivering it to fragments from tip to root. Howling like a pack of famished wolves, or like Taurus when he puts his muzzle to the earth and bellows defiance and the battle-cry to his adversary, the wind, surging around the mountains, drives the wild, angry clouds in furious eddies, and the dreary darkness of despair invests the scene. As if the floodgates of heaven were opened, the rain descends, and mighty torrents drench this little world. Slowly the clouds retire, the main body wheeling round some distant summit yet unvisited; the rear columns, following rapidly, keep up for a while a desultory pattering, and the sun, struggling through the vapory veil, shows that nature can smile through her tears! Rills and rivulets, foul with the débris of the forest, come noisily tumbling down. The Notch is now a broad and rapid river, the trees put on a brighter green, and all the scene looks smiling, fresh, and gay. The atmosphere is purified, the burning and suffocating heat is dispelled, and crowds issue forth to enjoy the refreshing breeze that has kissed the cleansed and purified summits of so many mountains, and fragrant with the odors of the great, still dripping wilderness!

Large and commodious houses have been erected here for the accommodation of the visitors at this place. As this is in no sense an agricultural section, and everything, including provisions, has to be brought here, and as transportation is both difficult and expensive, the charges at these houses are necessarily high; but the most sensible, as well as the most convenient, method of visiting these regions is for a party to take a shelter tent, cooking apparatus, etc., and, procuring supplies at the nearest possible point, camp out, and supply their table with trout and such game as is seasonable. Pitch the tent on a gentle, grassy slope, under the generous shade of grand old trees, arrange a comfortable bed of healthy, fragrant hemlock boughs, and live like a king. Uniting the duties of purveyor and cook, with a powerful and convenient appetite and plenty of fuel, anything that is wholesome is relished keenly. Thus with a small outlay one may see more and enjoy more, and be independent, going whithersoever curiosity or inclination prompts, returning at pleasure.

In this manner the interval snatched from toil and business may be profitably employed and heartily enjoyed. Exploring the sylvan retreats, the cool, secluded glens, the mysterious caverns, the deep, dark, trackless wood, scaling the slippery cliffs and towering pinnacles, surveying the mysteries of this mysterious territory, or exploring for others yet undiscovered, reading, refreshing and storing the mind from this illustrated and interesting and instructive page of the book of nature, is ample compensation for a year of toil. Mind and body together recuperate, a new strength of the system and of resolution is obtained, the heart is stouter, the courage stronger, a renewed and increased faith displaces doubts and misgivings; in short, a general regeneration or reconstruction (mental and physical) rewards, in addition to a pleasant season of rational enjoyment, those who try the experiment.

The Merrimack River has its source in the heart of the White Mountain region. Its head-waters are known as the east, middle, and west branches, which, uniting in the town of Woodstock, formerly Peeling, form at once a beautiful and important river. The east branch is much the most considerable, having its source some fifty miles above the confluence of the triangular tributaries, and receiving many accessions, it arrives here a splendid river of beautiful waters, and is the grand central figure, throughout its whole extent, of the great wild valley of the upper Merrimack, which comparatively few people know anything definite about, and fewer still have yet had the rare fortune to explore. The middle branch has its source in the Profile Lake, at the base of the Old Man's eternal throne. The western branch comes far up from the Moosilauke country in the wilderness of timber, where wild beasts maintain as yet almost undisputed sway, and all the tributaries of these three branches as well as themselves, great and small, the capillaries of the Merrimack, are all alive with trout.

Having descended from the great altitude where its source is located, and traced the romantic career through the wild and secluded valley until it has become a very considerable and beautiful river, let us proceed along the course of this resplendent stream, depicting its services and its value as it steadily moves forward on its journey to the sea.

CHAPTER III.

Franconia. — Lincoln. — Woodstock. — Thornton. — Game and Gunning. — Campton. — Holderness. — Samuel Livermore. — Plymouth. — Rumney. —Wentworth. — The Moosilaukes. — Bridgewater. — Squam Lakes and River. — New Hampton. — Bristol. — Newfound Lake and River. — Hill. — Smith's River. — Orange. — Andover. — Salisbury. — Daniel Webster. — Kearsarge. — Boscawen. — Franklin.

FRANCONIA, which supplies many streams to swell the Pemigewasset, is in many respects the most remarkable town in the United States, and for its great natural curiosities, its sublime scenery, its high mountains, it is extensively famed; indeed, it is believed to be in this country or any other, in this regard, without a rival.

"One of the greatest natural curiosities in the State is Profile Mountain in this place, near the road leading to Plymouth. This mountain rises very boldly a thousand feet high, and forms with its front of bare solid rock a perfect likeness of a human face." *

Formerly there was a large iron business done here, but it is now suspended. The New Hampshire Iron Factory Company was incorporated in 1805. The ore which was mined in Lisbon, three miles from the factory, yielded from fifty-six to sixty-three per cent. pure iron, and was thought to be the richest mine in the United States, and apparently inexhaustible. These works were actively operated for many years, and more than six hundred tons of hollow ware, stoves, etc., were turned out annually, requiring the consumption of more than three hundred thousand bushels of charcoal.

The principal business of the town now is, next to agriculture, to provide and care for the mountain visitors, who, it is estimated, arrive here at the rate of more than one hundred per day, exclusive of private conveyances, during the season of mountain travel.

Passengers reach here from New York, via Norwich, Worcester, Nashua, Concord, and Littleton, or by stage from Plymouth, or via Centre Harbor, Conway, and the Crawford House.

* Geography of New Hampshire.

As Franconia is the coolest locality in the United States in the summer, so it is in the winter; and whenever a severe cold term occurs, the telegraph invariably flashes the state of the thermometer at Franconia all over the land. Holton and Marquette, on Lake Superior, and St. Paul, Minnesota, often exhibit a great depression of temperature; but when looking for forty degrees below, where the temperature is of such intensity that the limbs of trees snap with a continuous rattle like a fusilade of small fire-arms, while the earth, and even the rocks, contract with reports resembling heavy siege artillery, by the terrible, frigid severity of the atmosphere, then it is that Franconia's bitter and appalling polar temperature figures almost invariably a trifle lower than any other locality.

Lincoln is the next town following the river to Franconia, and is very rough and mountainous. There being but a small portion of tillage land, the town is sparsely settled. Directly under the shadow of the great Franconia Peaks, it is subject to frosts late in spring and early in the autumn; consequently the cultivation of such products as are most affected by this cause is generally abandoned.

Kinsman's Mountain is so steep that it is seamed with slides which have furrowed deep channels in its sides from summit to base. The greater portion of the town is yet a wilderness, and bears, deer, and other smaller game abound.

In Woodstock, formerly Peeling, is the junction of the three branches of the Pemigewasset.

The general topographical appearance of this town is the same as Lincoln.

There are four or five large ponds, the streams from which afford numerous mill-privileges, and are generally used for saw-mills and such purposes, and furnish excellent trouting.

The great lumber companies of the lower Merrimack, whose head-quarters are at Lowell, have obtained an immense quantity of lumber from this town, often employing in the timber forests as many as a hundred and fifty men, who use this most useful and convenient river for the conveyance of the lumber. Some ten million feet are annually carried beyond the limits of the State.

There is a large cave here, sufficiently commodious to contain several hundred people, which is called the Ice House, from the fact that ice is obtained in it all through the summer. The walls of the cave are solid granite, and it communicates with several subterranean passages or apartments extending in various directions. The Grafton Mineral Springs here are strongly impregnated with sulphur and other minerals, and are resorted to by invalids who realize beneficial results. Blue, Cushman's, and Black Mountains are the highest elevations in the town.

On Moosilauke Brook there is a beautiful cascade, where the water glides over a smooth surface, or tumbles in a milky foam two hundred feet, remarkably picturesque and even sublime.

There is in Thornton, on Hill Brook, a magnificent waterfall. For thirty feet before reaching the final falls the descent is one foot in four, when it makes a perpendicular leap of forty-two feet. The town was originally granted to the Thorntons, Mathew, James, and Andrew; hence its name.

Mathew Thornton was born in Ireland, in 1714, and, like millions of his countrymen, preferring liberty in a foreign land to the most galling thraldom at home, however painful the struggle when compelled to make a choice, he came to the United States, and settled on the Merrimack, in Londonderry, as a practising physician. Bred in the traditional animosity of his people towards Great Britain, their oppressor, he espoused the cause of the colonies against the mother country with a heartiness and ability which proved him a hater of tyranny and England, and no mean foe for her to contend against. He was in all the stirring times preceding and during the Revolutionary struggle, a prominent actor; was with Sir William Pepperell in his expedition against Louisburg, in 1745, and was president of the first provincial convention, in 1775. In the following year he took his seat in the Continental Congress, and had the fortune to sign the roll which gave his name merited and imperishable renown, — the Declaration of Independence.

Thus Mathew Thornton was "one of the few immortal names that were not born to die."

He died in 1803, after a long life of devotion to the cause of self-government, full of honors, leaving an unstained character.

The north-eastern boundary of Thornton is the vast tract of ungranted land which comprises the territory of the Franconia and White Mountain region to the west and north, the Breton Woods, Hart's and Sawyer's locations to the east, extending southward to old

Chocorua Peak. It is the very heart of this magnificent wild that gives the Merrimack to the use and profit of man. In this great solitude, — which excels even the famous Adirondac country in the wildness and grandeur of its natural scenery, in the beauty of its forests, the number, form, and majestic proportions of its mountains, in the solemn, oppressive silence which pervades its whole extent, in the number of its cool, pure, pellucid springs, brooks, and ponds, which is so secure from the invasion of man that the sounds and sights of civilization are completely shut out from its extensive recesses, — there are game and fish worthy of the sportsman's attention, skill, and courage. The east branch, and, in fact, the whole vast network of waters about the source of the Merrimack, are bountifully stocked with trout, and the places which once knew the superb salmon, it is hoped and believed, will soon know him again. Each cold, transparent spring sends down the mountain-side its inimitably musical rill, which, meandering through the shaded reaches of the forests, pours the wealth of its generous flood into the cool, secluded pond, which, in turn, disgorges through circuitous channels to the swelling bosom of the Merrimack. Being a dense, luxuriant, unbroken forest, where the woodman's axe has never yet been heard, many parts of which the white man's foot has never traced, the fly can nowhere be used except in the Pemigewasset itself, or on some portions of some of its largest tributaries, and on the ponds. Here the fly, of a color best suited to the season, may be advantageously employed, and the best time to take the trout is from the moment there is a streak of day till sunrise or after, and again when the sun declines below the forests and the hills until dark. But the brooks may be successfully followed any portion of the day, yielding an abundance. The trout in all these waters are uniformly the best variety of salmon trout, varying in size of course; specimens being taken weighing from one-fourth to two and a half pounds and upwards, — the latter large enough to create the most nervous excitement and anxiety while the gorgeous prizes were being secured, and sufficiently large and luscious to satisfy any reasonable ambition or palate.

The wealth and productive capacity of the vast network of the upper waters of the Merrimack in the way of trout, invariably of the finest variety, is astonishing. For many years thousands of people have annually visited the White Mountain region; the gentlemen indulge, many of them, daily, as also many of the ladies, with a delicious enthusiasm, in the exciting and exhilarating sport of trout-fishing. Added to this, parties of experts are constantly arriving throughout the season from all parts of New England, and still the supply seems unlimited. A gentleman, in July of the present year, dropped his hook into one of the numerous slender tributaries of the Merrimack, a mere brook in Thornton, and within an hour had basketed nine of these princely fish, — the string weighing fifteen and a half pounds, an average of about one and three-fourths pounds. This is by no means a solitary case.

Since the exodus of the salmon, fly-fishing has almost fallen into disuse, and there are, probably, very few experts with the fly on these waters; however, no one can learn this art except with the rod in hand, as all the conditions differ at each season and fishing-ground. A correspondent of the "American Union," of Sept. 5th, 1868, — a paper which not unfrequently contains useful and interesting information on the higher order of field-sports, — gives some valuable practical hints, although he fully realizes and recognizes the total impracticability of theoretical fly-fishing. He says: —

" The art of throwing a long and light line can only be acquired by practice. The only rule I can give is, to let it go its full length out behind you before you switch it forward, and, of course, study the wind. This grand fundamental point of skill being reached, — namely, eluding the eyesight of the trout, who all lie with their heads up the stream, — the next question the learner should ask of his instructor is, on what spot of the water shall I allow my line to light? It is only to be arrived at inch by inch, and only to be fully mastered by years of practice and observation. The lure must light true to an inch. Beginners fancy that trout are scattered in a random way over the water, and that they are just as likely to encounter a trout in one place as in another. There is, perhaps, a prejudice in favor of deep pools, which are generally the most difficult places in which to kill trouts, from the quantity of line which you must expose in order to reach them. The strongest trout of the neighborhood selects the best spot for feeding, and so soon grows bigger and stronger still, while his poorer neighbors are struggling for the means of bare existence, — just as the rich man in this world, by strength of funds, tends to grow rich, while the poor still remain poor.

"Now the question is, how to discover the spot which the potentate will have chosen. He will have had an eye to the likeliest bit for securing flies and worms as they come down the stream, or drop off trees and bushes. He will have also given careful consideration for his bodily ease, not liking, generally speaking, to be bored by a strong current. But I could give the reader more insight into this question in an hour at the river's side than by any amount of writing. The still water at the neck of what fishers call the 'stream' — that is, the rough, rapid water as distinguished from the still, or pools — is a favorite haunt. In summer, when the trout are feeding, the edges of pools are favorite resorts. There the trout are often lying in hundreds, digging with their snouts into the banks for worms.

"While there are, certain casts that are nearly always good, the general feeding-ground varies with the weather and season; hence the great difficulty of arriving at a thorough comprehension of this most important point. How important, any one may understand when told that, when really feeding, the trout almost all fly to the same character of water, and while the skilled angler is pulling them out at every cast, the unskilled one is employed in vainly thrashing water devoid of a single fin. The meal being finished, or the shower of flies which induced it having left the water, the trout then return to the deeps to ruminate and digest, and while this process is going on he will be a cunning angler indeed who will induce them to take his lure.

"In warm weather in summer, when the trout begin to feed, they all leave the deep, and come into the shallowest water, — often into water so shallow that it hardly covers the backs of the large ones. The reason of their doing this is, that in the summer flies hover near the surface of the river, but seldom fall on to it, as they do in spring, when they are weaker; so that to catch them by a leap the trout must lie near them. At no time are good trout so little shy, or so greedy, as when they are lying in water so shallow that none but the skilful angler ever thinks of throwing a line into it.

"The next point to which I will advert is precision of aim. Suppose you know, as the cunning angler does, almost to half an inch, where the maw of his destined victim is placed during his feeding-time, and that you understand how to elude his watchful eye, there is still something else to be done before you are successful in a well-fished river. If your lure falls lightly, and, consequently, naturally, within an inch or so of the trout, his instinct will lead him instantly to snap at it. If the lure be a bait, he will proceed to gorge after catching it, provided he does not feel the line tighten, or see you, when he will instantly let go his hold if he can. An artificial fly, however, he will reject the instant he touches it; hence the importance of quick, natural, and well-trained eyesight, in this branch of trout-angling, to enable you to strike the instant he touches the hook, — that is, before you have felt anything. If your first intimation of the trout's attack in fly-fishing comes from your sense of touch, as is the case with ninety-nine anglers out of a hundred, you need not strike at all, — he has either hooked himself, or else he is gone.

"Suppose, however, that you bungle your cast, and the lure falls five or six inches from the trout, he will then most likely dart at it, but the odds are a hundred to one that he will become suspicious, and turn tail before reaching it. He will have had time to take stock of the line, the shadow of the rod, and very likely of yourself. I have, hundreds of times, on making a false throw, seen the trout dart from his lair, with the view of seizing the lure, then catch sight of the line, and wheel back, quick as lightning, to his hiding-place. In fishing-pools, where the trout are digging in the banks for worms, the most absolute precision of throw is requisite; because the water being still, and probably six or eight inches deep about the edge, — if deeper than this you will do nothing, — the line will certainly be seen; and if you give the trout a second for reflection, he will infallibly say, 'No, thank you,' to your proffered morsel.

"When the streams are low and clear, there is nothing more important than fine tackle; the finest, *roundest* gut that can be obtained to be used, and the joinings should be firmly and neatly knotted. Never use loops. A loop is an abomination in the eyes of a well-fed lowland trout. It is undoubtedly curious that when a trout will confidently seize a minnow, or a worm, with two or three hooks sticking through it, the sight of a bit of coarse or flat gut will terrify him out of his wits, or, to speak more correctly, perhaps, will terrify him into them. It is certainly true, however, that line, and not hooks, is what he dreads. For this, and for another reason, the less line you throw into the water the better. The other reason is, that, as a general rule, the less line you have in the water, the more natural will be the movement of your lure. Suppose the stream, at the spot where it alights, to be running at three miles an hour, and that the upper part of your line falls where it is running at six miles an hour, it is plain that your hook will be dragged down in an unnatural, and, therefore, suspicious manner."

Game, of the larger varieties, is plenty in this great wilderness, but on account of the almost inaccessible mountains, the deep gorges, and the secluded glens, is not so easily taken as in some other parts of the country. Margins of lakes, ponds, and watercourses are pathed by the deer, bear, and other animals. Deer, being mild and harmless, have many enemies, not the least of which is man. The natural defence of the deer against a world of foes is fleetness of foot; but the deep snows of winter in these high, northern latitudes sadly interfere with the rapidity of his locomotion, and he falls a victim to the appetite of many voracious brutes. At this season of the year the habit of the deer is to "yard," as it is called. Selecting, if possible, an extensive growth of low hemlock, they assemble in small numbers, and, by moving about to browse the branches, keep the snow trodden. If the depth of snow permits, they move as soon as they have trimmed this yard closely, but a deep snow and a hard, thick crust are fatal if their retreat is discovered by man or hungry beasts; their weight and sharp hoofs cutting through at every bound the soft, deep snow underneath, leaving no solid basis to spring from. A man, living on the eastern border of the ungranted land, showed the party where he captured seven in this manner, early in the day, a mile only from his house, the poor creatures being entirely helpless. Encamped in the solitary depths of this wide wilderness, with a few choice spirits, near some tumbling torrent of sweet waters, whose liquid music, more delicious than a viol, harp, or grand piano, lulls one to repose, far away from the bustle, din, and strife of the great city; or with a raft of logs — a pleasant but primitive mode of navigation — float on the tranquil surface of the mountain pond, and see the blue sky and fleecy clouds gaze down upon themselves in its clear depths, or the bright stars, like glittering eyes of countless mermaids, flashing up from the apparently unfathomable deep to snuff the sweet zephyrs freighted with health and the fragrance of the forests, to seek recreation, comfort, and sport in these grandly furnished apartments in the great temple of nature, to feast on trout and mountain grouse, concluding the day's enjoyment with a sweet and pleasant meerschaum, is a luxury fit for a king, or any other man.

As the shades of night approach, and the lengthening shadows give admonition that the sun is sinking fast behind the western hills, and night and darkness creep slowly on, one reflects that he is almost helpless in this great wildwood, surrounded by fierce beasts and an impenetrable gloom, and a kind of pleasant or not disagreeable loneliness steals over the feelings. The city — where, in gilded palaces and gorgeous costume vice reigns triumphant now; the interminable lines of brilliant gas-lights, where the struggle and turmoil for the almighty dollar is now progressing; where the rush for the bulletins and the news is so furious; where all the elements, good and bad, of the human character are wrought to the highest pitch and strained to the utmost tension for good or for evil, — in this regard almost a neutrality, — comes rushing in a blaze of artificial glory to the view, and the mind involuntarily recoils, and the feeling is, "I am safer here than there, and better, socially, morally, and physically." The blazing camp-fire now throws out its lurid glare, and the countless trunks of giant trees, stretching into the darkness beyond the farthest *scintilla* of illumination, seem columns in the giant's causeway of this wilderness, or pendent stalactites in this mammoth cave of overarching foliage. Cries of wild beasts are heard about the neighborhood, attracted hither, no doubt, by the glaring fire, or the flavor of savory cooking; but they are not dangerous, the most serious consequences of their proximity being the electric shock experienced at their terrifying cries, the wild bound of the heart, and then its pause, the awful terror that thrills the soul, and the apparently intoxicated reeling of the trees. Only a few years since a wildcat was killed near Thornton, which was fierce and powerful.

The party around the camp-fire received a visit from one of the buff, bobtailed species, whose weight was estimated at between thirty and forty pounds; and less than ten years ago, one of the monster brutes of this great wilderness, which had strayed away from his haunt, was killed in Lee, a monster panther weighing one hundred and ninety-eight pounds.

Mountain grouse are very plenty and delicious; sable and mink are profitably trapped, and Canadian otter are frequently taken; but the fox, that sly, cunning rascal, avoids these solitudes, and is rarely found far from the habitations of men. Reynard has an unconquerable appetite for poultry; consequently the rugged cliffs, and bleak mountains, and desert wastes of this unsettled wilderness have no charms for him; his tastes run in the direction of open and cultivated places, where the gospel is preached and chickens are raised. Thus, Sunday, under that patron of morality, the church, while people assemble for worship, and under "that patroness of rogues," the moon, when people are at rest, this sly old reprobate prowls for his prey.

The raccoon, or, as he is commonly called, the coon, abounds in all the forests, especially the nut-bearing woods, which border the Merrimack and its branches, and is an animal worthy of some attention. Except in one single particular, he is perfectly harmless, and that is a decided taste for Indian corn, — at no time, however, except when it is in the "milk." His depredations are annoying to the farmer, as his usual practice is to break down an ear, strip the husk from one side, take a mouthful, and proceed to break down another. Chestnut, walnut, and oak forests are his usual haunts, these nuts being his favorite food; he, however, loves frogs, like a Frenchman, and the skilled coon-hunter directs his course to a stream passing through nut-bearing woods, where the coon expects to luxuriate on nuts and frogs' legs. He is never hunted in the daytime, being a nocturnal rambler. The most favorable time for coon-hunting is regarded by experts as during the October moon, as at that time the young ones are well grown and fat. Proceeding to the vicinity of a stream, the margin of which is the most "likely" place to "pick up" a fresh track, giving time for the animal to wander some distance from his den, in a ledge or hollow tree, the hunter must have a stanch, well-trained coon dog, or nothing can be accomplished, as tracks of other animals bewilder an unreliable dog. Towards nine o'clock the dogs are let out, and if there is a coon in the vicinity he is soon treed. As many as three are frequently taken, and sometimes five from a single tree. At this season he is uniformly fat, and is excellent eating, making a superb fricassee, care being taken, in dressing, to remove the "kernels" or fatty secretions directly under the fore legs, which secures the meat from that strong musky flavor, which otherwise renders it entirely unpalatable. In 1865, E. L. Chapman, Esq., proprietor of Chapman's Hotel, Epping. N. H., shot a fine specimen, weighing thirty-three pounds. There is a true nobility of sport in successful coon-hunting not realized or appreciated except by those possessing a genuine taste for it, and who are generally philosophers in some sense, naturalists in another, and invariably splendid fellows.

The woodcock, in its annual migration northward, arrives in the Merrimack Valley early in the spring, advancing with the season, often before, or more rapidly than the tardy approach of mild weather would seem to warrant. It breeds early, and its eggs are sometimes seen deposited on some dry, protected spot, contemporaneous with surrounding patches of snow. It is, however, a hardy bird, and these unfavorable conditions neither interfere with its habits nor its regular semi-annual hegira. It is exclusively a ground bird, — never alighting on trees, — and its favorite feeding-grounds are dry alder-swamps, corn-fields, and such other localities as supply food, and are easy to penetrate with its long, robust beak. The gunning season commences the first of August, at which time the young are nearly full-grown, and the birds are uniformly in excellent condition, and continues until the ground is closed by frost, when the birds retire to winter-quarters in a more congenial clime. In woodcock-shooting, as experienced sportsmen well understand, it is policy for two to gun together, and, seeking the most approved feeding grounds, with a pair of stanch and well-trained setter or pointer dogs, the party not unfrequently bag from two to two and a half dozens of these fine birds, besides generally a small assortment of grouse, pigeons, plover, gray squirrels, etc., etc. Woodcock-shooting is a most delightful and exciting sport, there being, perhaps, nothing in New England, in the way of gunning (for birds), away from the sea-shore, which can equal it. The flesh of this bird is esteemed a rare delicacy; the sport is exciting and exhilarating, and the taste and skill of those who most indulge in it designate them as the true nobility of sportsmen. Another feature, attractive to many, is that as these birds are eagerly sought in the market, at five dollars and upwards per dozen, the skilful sportsman may make it as profitable as pleasant.

Upland plover, a bird of fine flavor, is also frequently met with, not, however, in such numbers as to make the sport, like that of woodcock-shooting, a specialty.

Gray squirrels, and occasionally black, are numerous throughout all the territory watered by the Merrimack and its tributaries. Their especial haunts are precisely those of the raccoon, their food being similar, — acorns, nuts, and corn. Oak, walnut, and chestnut woods and the borders of cultivated sections are the favorite resorts of the gray squirrel. Successful gunning for this animal requires experience and much prudence, as he is wary, active, and cunning. There are two principal methods of squirrel-hunting, both practised in turn by the knowing ones. The first is this : each individual sportsman repairs alone to the nut-bearing forests before the frost has nipped the leaves and thinned the foliage ; reaching an approved locality, the sportsman at once stations himself in a position to attract the least attention, while he, at the same time, may obtain a good view of the surrounding trees and overhanging branches. Having secured an advantageous location, he maintains a stolid immobility and silence; the unsuspecting victim approaches within range, when the unerring "double-barrel" speaks sharp and quick, and he comes down. Having bagged several in that vicinity, and spread consternation by the repeated report of his piece, the sportsman moves to some other locality, and operates in the same manner as before. This method is known to sportsmen as the "still hunt." When autumn and frost have stripped the tall chestnuts and oaks of their summer dress, the sportsman, with a trusty squirrel dog, again repairs to the same forests. At this time much depends on the dog, who is not only expected to "tree" the animal, but to indicate, unerringly, precisely what tree the squirrel is on. There is no section of the Merrimack Valley in New Hampshire, even to its extreme head-waters, where gray squirrels are not plentiful, and there is no season when the experienced sportsman may not enjoy a pleasant and profitable day's gunning for this description of game. The flesh of the squirrel is esteemed by many as a luxury, and is certainly a savory delicacy, and its fur is also of some value. Hares, in large numbers, abound in all except the mountain forests of the Merrimack

Valley. The gunning season commences about Christmas time, and continues until "March Meeting;" it is prosecuted to the best advantage in sheltered swamps, where, usually, as many as is desirable may be obtained. They are "run" in the same manner as the fox, describing, however, circles of less diameter, though, perhaps, more eccentric. The best hare dog is a cross (half hound), as it reduces the natural disposition to range, or leave the track for that of the fox or larger game. The hare is harmless to the products of the farm, changes his grayish-brown coat in the fall for a suit of spotless white, and is hunted exclusively for his flesh, which is regarded by most people as excellent food.

The territory which is now included in the towns of Campton and Rumney was purchased by a party of actual settlers. While making the necessary surveys and other arrangements for permanent occupation, they constructed and dwelt in a temporary camp; hence the name. Mad River, one of the tributaries of the Pemigewasset, is a rapid, and, at times of high water, a furious stream. Its source is in the mountains, to the north-east in the ungranted territory, and unites with the Pemigewasset in this town, and is well known to sportsmen. Squam Mountain is in this town. The view from Mt. Prospect is splendid; except, perhaps, from Red Hill in Moultonborough, it is believed the view of Lake Winnipesaukee is the finest to be obtained anywhere, and a large portion of the southern part of the State may be surveyed from this fine mountain. It is often visited by excursion parties, being less than five miles from the Pemigewasset House in Plymouth. Mad and two other small rivers fall into the Pemigewasset on the east or left bank. Plumbago is found in large quantities, and of a superior quality. Iron and other minerals are also found.

Livermore's Falls are in the Pemigewasset in this town; they exhibit evidence of volcanic formation, and considerable fragments of lava have been found. In 1752, four men, namely, John and William Stark, of Derryfield, now Manchester; David Stinson, of Londonderry, and Amos Eastman, of Pennacook, were trapping on Baker's River, in what is now Rumney, but was a portion of the Campton grant of Captain Spencer; they were discovered by a party of Indians, who resolved to capture them. Thinking they observed Indian signs, the trappers determined to take up their traps and leave, as they had nearly three thousand dollars' worth of furs and other property to secure. Accordingly, on the 28th of April, while in the act of taking up a trap, John Stark was captured. William Stark made his escape, and, while attempting to do the same, David Stinson was killed and scalped. Eastman was captured, and John Stark and Eastman were taken as captives to St. Francis. As is the custom with the Indians, the prisoners, on arriving, were compelled to run the gauntlet, between two files of Indians. Eastman went first, and was horribly bruised and kicked, they having previously been instructed to repeat a sentence in Indian, which signified that they would beat all the young men of the tribe. Stark, instead of repeating this, declared, in a loud voice, he would kiss all the young women; and, being athletic, the few buffets he received were returned with such earnestness and interest, that some were knocked down, and others, alarmed at his boldness and prowess, were fain to allow him a peaceable transit.

Stark became a great favorite with the Indians, and, both being ransomed in the July following, they required double the price for Stark that they did for Eastman. John Stark lived to be the hero of the important battle of Bennington, and won unfading laurels.

Great and Little Squam Lakes are located principally in Holderness. Squam River, which discharges the waters of these two lakes, falls into the Pemigewasset near the south-west angle of the town, on its east bank, immediately below Bridgewater Falls.

The great corporations along the lower Merrimack have provided artificial means to draw the whole surface of these two lakes, reducing them several feet when the dry season affects the natural volume of the river. Thus the Winnepesaukee and Squam are a treasury of waters from which to draw a never-failing supply to keep the wheels and spindles moving.

There are excellent mill privileges here and shops of various kinds, saw and grain mills, paper-mills, and some cloth manufacturing, shoe-peg factories, etc. The town was granted to John Shepard and others, in 1751. Hon. Samuel Livermore settled in this town in 1765, and became proprietor of a large part of its territory. He was a leading and prominent man in this section, and for several years was attorney-general for the crown. He was a judge, and for eight years United States Senator, from 1793 to 1801.

He was a man of good abilities, dignified, austere, and possessed great self-esteem. When in Congress, like some of his successors, he sometimes ventured into deep water, and was rescued by the exertions of friends, or perhaps by the forbearance and magnanimity of his antagonist.

Mr. William Shepard, a native of Holderness, and a descendant of John Shepard, the original settler, who died recently in Londonderry at an advanced age, and who accompanied Judge Livermore during several sessions of Congress, related many anecdotes bearing on this point.

Mr. Shepard at that time was above his majority, consequently old enough to retain an accurate recollection of the personal appearance of many of the principal men of that time, such as Generals Washington, Hamilton, the elder Adams, Jefferson, Robert Goodlow Harper, Judge Cooper, John Randolph, and others, as well as the part which Judge Livermore took in the stirring political drama which these historic actors were performing at that period. As Mr. Shepard retained possession of his faculties in a remarkable degree to his last days, the fund of reminiscence, incidents, and anecdotes which came into his possession by personal observation was very extensive and highly entertaining, and it may also be observed that few ordinary men ever enjoyed the rare fortune of meeting with so many illustrious men as fell to the lot of Mr. Shepard. Judge Livermore died in Holderness, where he had resided uninterruptedly for nearly forty years, in 1803.

Judge Samuel Livermore was an educated, courteous, and high-toned gentleman, but conceited, self-willed, and arbitrary. His habitation was in the clouds, and he never descended to mix or mingle with ordinary mortals. He was occasionally admired, generally respected, sometimes feared, but never loved. He possessed but little sympathy with his kind, and, leaving no enduring monument of positive greatness, was soon forgotten when the grave closed over him. In his day he was, undoubtedly, the great man of New Hampshire, but his local pre-eminence or prominence was, perhaps, due more to the scarcity of men above mediocrity in the State than to his intrinsic merits in that direction, intellectually considered. His natural force of character and self-will gave him a position in the front rank of his contemporaries; still, when he had passed away, the record of his career proved a meagre theme for history. Though a strong, sagacious man, his bitter prejudices and arbitrary spirit made him unreliable in party and public relations, and, although respected and counselled with by such men as Hamilton, and other of the foremost men of his time, yet he was of such brittle material, and required such careful handling, that he was probably not a coveted acquisition to any party, or in forwarding any scheme. In his congressional career he was active, watchful, and efficient in the practical details of legislation, but never grappled with the great questions of national policy that divided men and parties and brought out the herculean efforts of a race of giants; he also had the prudence, generally, to avoid the contact in the forum of the great men, as he did of important measures. As a judge, he loved justice and administered it impartially, without either prejudice or sympathy. He was not a great lawyer or great judge, yet he was honest, impartial, and scrupulously faithful in the performance of all his judicial duties.

Plymouth and Haverhill are the shires of Grafton County. Baker's River, which rises in Moosilauke Mountain, effects a junction with the Pemigewasset on its west or right bank in Plymouth. The river is named in honor of Capt. Baker, of Haverhill, Mass., who attacked the Indian settlement at its mouth, routed them, and captured considerable booty, furs, etc.

The Pemigewasset House, in Plymouth, is one of the largest and most famous hotels in the State. There is a court-house here, and an excellent seminary, which is in a flourishing condition.

From this place there is a good stage road through the valley of the Pemigewasset to Franconia Notch, a distance of about thirty miles.

Rumney, north of Plymouth, on Baker's River, is a beautiful village at the mouth of a large stream called Stinson's Brook, the outlet of Stinson's Pond, so called in honor of David Stinson, of Londonderry, who was killed here by the Indians. Stinson's Mountain is also in this town.

From the village the road to Ellsworth through Doetown, so called, winds up a long, steep hill. Arriving at the summit, which may be one and a half or two miles from the village, the most delightful landscape bursts upon the view. Looking down the long, wearying declivity just ascended, the neat village of white cottages nestles on the broad green lawn which stretches away to the river, meandering like a wavy belt of silver, with its picturesque, covered bridge. Just beyond, and parallel with the river, the handsome dwellings, built along the hill-side, terraced and hedged and ornamented with fruit and shade trees; to the left is the hill and dale down to Plymouth and the Pemigewasset valley; a little farther left the peaks of the grim and bare Sandwich, and the grand and densely wooded Squam Mountains; to the right, Stinson's Pond, a long, narrow stretch of water, with hills to shut it in from the west, while turning to the north is seen a promiscuous assembly of mountains, rising high and still higher, — Moosilauke towering above his fellows, Lafayette in the distance, and Mt. Washington in the rear rank, looking cool and pure over all other heads.

There are grander views than this, more extensive, affording a greater amount and variety of land and water scenery; but for unpretending loveliness, for a combination peculiar and rare, giving the whole vista an indescribable charm, there are few places to equal this, none to excel.

Still further up the river is Wentworth. There being a fall of twenty feet, the river here affords good water-power.

Moosilauke Mountain is the highest elevation west of the White and Franconia ranges. The north peak is four thousand seven hundred feet high, and is a conspicuous object from all the country round, far into Vermont. There are inexhaustible quarries of limestone here, and along Baker's River and about the Moosilauke, minerals in great variety abound. Plumbago, tin, galena, lead, iron, tremolite, black blende, and crystallized epidote, are found.

The beautiful Pemigewasset, having received at Plymouth the addition to its steadily increasing volume of Baker's River, sweeps along the eastern boundary of Bridgewater, receiving, as it passes the succeeding towns, the copious additions of Newfound and Smith's Rivers, from the west or right bank, and Squam on the east or left. Squam River has its source in Squam Lakes (Great and Little), which extend into several towns to the east of the Pemigewasset and near the unsettled territory.

Newfound River has its source in the large lake of the same

name, which extends into several towns to the west of the Pemigewasset, and is admirably adapted to manufacturing purposes. Indeed, it is believed that Bristol will soon be a famous manufacturing town.

There was an academical and theological institution established in New Hampton nearly fifty years ago, under the patronage of the Baptists; also a female seminary, justly famed for the completeness of its instruction and the picturesque beauty of surrounding scenery. These institutions were removed to Vermont some time since, but there is an academy where mental and physical improvement may be obtained.

There is a peculiar spring on a hill in this town which is never affected by wet or dry weather, maintaining an even tenor and issuing a stream sufficiently large to carry mills located along the course of its waters. There is also a peculiarly formed hill, called the "Cone," being remarkably conical, and noticeable from the surrounding country many miles distant. The view from its summit is most charming.

New Hampton village is nearly one mile in length, is very level, parallel with, and about a mile from, the Pemigewasset. It is pretty, pleasant, and quiet, and, only for the beautiful school-girls, the stillness would be absolutely painful.

A very noticeable feature of the village is a row of nine large trees, directly in front of the hotel, the middle one being the tallest, and those on either side gradually and proportionately declining in height, presenting the appearance of having been shaped by the designing hand of man. This, however, is not the fact, it being the result of natural growth; and, as the trees are very near together, the branches interlace so closely that it appears only a single monster tree-top. This unique feature attracts the notice of strangers perhaps as much as the fine school-buildings, or the extensive and most fragrant flower-garden adjoining.

Directly at the bridge, on the highway from New Hampton to Bristol, is the head of the Bristol Falls. From this point to the mouth of the Newfound, — some eight miles by the course of the river, — the fall is continuous, though not rapid, until within a short distance of its foot. The topography of the adjacent land is very favorable for canals and factories, the chance and the material for constructing a dam is excellent, and the perpendicular fall is more than thirty feet.

The population of Bristol is estimated at fully two thousand, having doubled, both in population and resources, since the last general census. Newfound River, the outlet of the lake of the same name, passes directly through the entire length of the village, which may be said to extend already nearly the whole course of the stream; and the time is not far distant when the village will cluster along its whole extent. This river is, perhaps, one of the most remarkable streams in the State, being but two miles in its entire extent, and the liveliest two miles of river to be found anywhere. The total fall is upward of two hundred feet, one hundred of which is comprised within fifty rods of its confluence with the Pemigewasset, the latter stream sweeping the eastern boundary of the town. There is no space of twenty rods on this stream but is an eligible and efficient water-power, and the time cannot be far distant when the practical manufacturer and mechanic will occupy and operate it. The power already appropriated is as follows: Merrimack Hosiery, D. H. Rice, agent, produces some twenty-five hundred dozen pairs per month, taking the raw material and turning out the perfect article.

Crosby's bedstead manufactory turns out some two hundred bedsteads per week, which are sent south and west, many finding their way to the distant Pacific shore, — California and Oregon taking a large supply.

White's two tanneries, both on a very extensive scale; D. P. Alexander & Co.'s machine shop, which turns out a considerable quantity of machinery for various purposes: Lovejoy & Dow's carriage manufactory, where all kinds of vehicles are made, from the light and pleasant sun-down to a six-horse stage-coach; Crawford & Locke's flouring and grist mill, with a capacity of three hundred barrels per week, besides grinding twenty thousand bushels of western corn per annum and an extensive accommodation business of a local character. All who have the good fortune to patronize this firm gainsay the old adage, "You cannot have your cake and eat it too," and their patrons, to say the least, are extremely well-bre(a)d.

These mills are operated under the efficient supervision of George T. Crawford, Esq., a gentleman whose qualifications and business capabilities are demonstrated by the gratifying success which attends the transactions of the firm; nor is his ability exhausted by the business of which he has the managing control. His information is varied, general, and extensive, and his affability and explicit way of imparting information to those who apply to him is pleasant and refreshing. The business has been conducted by him for the past two years on the tide of general success, and this firm prides itself on supplying the New Hampshire market with the first new flour of the season.

Jordan's planing-mill and job works, — every variety of building material prepared for immediate use; Musgrove's wool-pulling, knit and flannel under-garments for gentlemen's wear; J. C. Draper & Co., Plymouth gloves, one of the most extensive and flourishing establishments of the kind in the State; D. Mason's straw-board mill, which finds a ready market in Lynn and Haverhill; Holden & Co.'s fancy flannel, employing some forty hands; B. F. Perkins, straw-board mills; Dow & Co., coarse woollens; the Lake Company's extensive lumber works, located directly at the foot of the lake, owned and operated by the large companies on the lower Merrimack.

In the village there are fourteen stores, two hotels, three churches, — Methodist, Baptist, and Congregationalist. The facilities for education, as in most other enterprising communities, are excellent. The Union school-house, completed in 1867, at a cost of twenty thousand dollars, is a fine structure, located on a beautiful eminence overlooking the entire village. It is a graded school, and accommodates the pupils in the district comprising the village proper.

Bristol is the natural and real business centre for the surrounding country, including the towns of Hebron, Groton, Bridgewater, Alexandria, Hill, and New Hampton. In addition to the immense water-power of the Newfound, a small portion of which is yet developed, the Bristol Falls, in the Pemigewasset, are available for an immense manufacturing production. Excepting the diminished volume of water, it is believed the capacity is not inferior to some of the largest places on the river, the fall being sufficient to allow of using the water over, and the facility for damming very convenient.

Newfound Lake is a magnificent sheet of water to look upon, but this by no means comprehends its value; indeed it is the smallest portion of its great merit. It is nine miles in extreme length, and about three in average breadth. The great land and water power companies of the lower Merrimack control this power, and, as in the case of Winnipesaukee and Squam, they have provided artificial means to draw this extensive reservoir to the depth of six feet of its entire surface, so that in the dryest seasons the stream affords a maximum supply for all the mills and shops along the course of its outlet.

The little elegant steamer Pioneer, Capt. Geo. W. Dow, plies on the lake, affording those delightful moonlight and daylight excursions which the citizen and tourist enjoy so highly. The new and spacious hotel of Mr. O. K. Bucklin is a marked and prominent feature of this delightful village. It stands at the head of the square; is four stories in height, and has all the modern improvements attainable in this section.

The village is romantically located in a basin, surrounded by a nearly circular wall of very high hills, whose peaks, some ten in number, rise like a line of silent sentinels to guard and protect the peaceful community, whose neat white houses resemble pieces of frosted cake in a huge tureen.

Newfound River tumbles noisily through the centre of the town, repeatedly leaping many feet at a single bound, and rushing in foaming cascades and rapid current with headlong speed until it is lost in the embrace of its larger brother. The resplendent Pemigewasset, washing the eastern shore of the village, diversifies the scenery of the thrifty and quiet town of Bristol to suit the tastes of all.

Any one desiring to escape the heat, dust, and confusion of the large cities will find this village accessible by the Concord and Bristol Branch Railroads, the latter passing through a country affording some of the most superb views. Through the deep-green luxuriant foliage, the traveller constantly catches glimpses of the sparkling waters of the Merrimack, with the long high range of heavily wooded sombre hills behind, while the broad, fertile intervals, diversified by fields of waving grass and grain, and the stately spires of extensive and luxuriant cornfields, afford a remarkably picturesque and charming view, diverting the unpleasant tedium of the usual railroad experience, and making Bristol a pleasant excursion, and Bucklin's Hotel just the place to stop at on arriving there.

A new and important enterprise which is destined to add largely to the population, wealth, and consequence of this town is being prosecuted with a reasonable prospect of a successful consummation. Reference is made to the new New Hampshire Central Railroad. Time is money, and the surest method of saving money is to save time, and to do this a direct line is the most rational and reasonable course to pursue. In order to accomplish this desirable result it has been found that the completion of a link in the great chain of travel from west to east is necessary, from Portland, Me., to Danbury, N. H., a distance of ninety miles, — a short break in the continuous chain through the State of Vermont of forty-six miles. When this is completed the distance from Portland to Chicago will be only one thousand fifty-four miles, while by any other route or proposed route it is one thousand one hundred and forty-six miles.

This route secures not only a very important diminution of the distance, but must certainly prove of value to every town through which it passes, as well as other towns within the reach of its influence. This line proposes to pass directly through this village, uniting with the Bristol branch, and effecting a junction with the Northern at Danbury, nine miles west of Bristol, giving this town the benefit of the local and through business alike.

There is in Bristol a flourishing lodge of the ancient brotherhood of the mystic tie. Union Lodge, No. 79, composed of about one hundred members, good men and true, embracing the leading citizens of this and surrounding towns; Moses H. Merrow, master.

Smith's River is the northern boundary of Hill, separating it from Bristol and falling into the Pemigewasset. The town was settled by people from Chester, Rockingham County, and was originally called New Chester. It received its present name in 1837. Ragged Mountain, whose altitude is but little inferior to Kearsarge, is in this town.

The Blackwater, a considerable tributary of the Contoocook, runs through the town parallel with the Merrimack, sometimes not more than four or five miles from it, passing through several towns and falling into the Contoocook at Hopkinton. In its course the Blackwater passes through the town of Andover, where repose the remains of the celebrated magician Potter. This man was one of the first, as he certainly was one of the most famous, of the practitioners of the black art.

Orange, where the source of Smith's River is located, is a small, poor town, limited in its agricultural capacity, but remarkable for the variety of its mineral productions, — iron and lead, yellow ochre in large quantities, superior in quality to the imported article, a conglomerate of chalk and magnesia, and from a pond is taken a mineral paint resembling a spruce yellow.

On the height of land which sheds its waters into the Merrimack on the east, and the Connecticut on the west, are a number of deep, circular holes worn in the solid rock, the largest of which, from its great depth and regularity of form. is called the "well." It is eleven feet in its perpendicular depth, and the stones found in it were rounded, smooth, and polished, indicating a powerful and long-continued action of water. The position of these wells is one thousand feet above the waters of the Merrimack and Connecticut; the rock is very hard, and in many places on its surface occur the marks and scars and scratches spoken of as evidence of the ancient drift epoch in the existence, or before the existence, of this continent.

Smith's River has its source near what is called Poverty Ridge, a ridge of sterile territory, extending north and south nearly the whole extent of Grafton County, being the water-shed between the Connecticut and upper waters of the Merrimack.

Salisbury is a very ancient town. It was originally granted by Massachusetts, and was called Bakerstown. It was regranted afterwards by the Masonian proprietors, and was known by the name of Stevenstown, — this was in 1749. It was incorporated in 1763 by New Hampshire under its present name. Until the formation of the town of Franklin in 1828 it extended eastward to the Merrimack, and its area was large.

In 1819, a huge mass of earth and rocks of many thousand tons weight, became detached from the southern declivity of Bald Hill, and was precipitated with overwhelming force into the valley below, bearing down every obstacle in its resistless career for a distance of nearly fifty rods, and four to five in breadth.

The Indians in their predatory incursions paid especial attention to this town; the early settlers suffering severely from their depredations and cruelty. In May, 1753, Nathaniel Maloon, his wife, and three children were captured and taken to Montreal, where himself and wife were sold, the children being retained among the Indians; one of them made his way home, arriving after an absence of nine years. In August of the same year, Mrs. Call was killed, and two men named Barber and Scribner made captives and taken to Canada.

Salisbury was settled in the year 1750 by John and Ebenezer Webster, and others. Ebenezer Webster, one of the earliest settlers, was an ardent patriot of the Revolution. and held many offices of honor and trust, but was most distinguished as being the father of Daniel Webster, — a name to which any title is unbecoming and superfluous; a familiar and honored name; in a long line of illustrious sons the noblest Roman, the most majestic, the greatest genius of them all. Daniel Webster was born in Salisbury, New Hampshire, 1782. Here he first breathed the atmosphere, and rested his eyes on the beautiful in nature spread out around his birthplace like a broad and magnificent panorama, embracing every variety of scenery, from the frowning mountain to the bright rolling river. Here he received his first impressions, his early training, and laid the foundation, broad and deep, of that magnificent and comprehensive intellect which made him in his time the peerless Webster. Reared in the midst of natural scenes like these, it may not be surprising that his youth was somewhat imaginative and poetical, nor is it more a wonder that his first efforts developed a mind at once compact and masculine. He started into manhood with an intellect, broad, fertile, and productive. Irrigated by a steady, pure, genuine patriotism, his great mind grasped his whole country, and his anxiety for the perpetuity of her institutions and his race was his noblest emotion. He despised fanatics with all the ardor of a generous mind, while he profoundly pitied them, and from whatever quarter they came he never lost an opportunity to give them a scathing and wholesome rebuke.

He graduated at Dartmouth College in the class of 1801, and commenced at once the practice of law. In 1812, though only thirty years of age, and as yet unknown to fame, he was elected to Congress, and there stood among the foremost in ability and in opposition to the war and the administration. He held all the high offices in the government, except the presidency, and most distinguished himself in the highest and most responsible.

That his political views did not obtain in the country is no evidence against his transcendent genius; on the contrary, the most dogmatical of his opponents were amazed at the consummate ability with which he sustained and urged what they considered a poor cause, declaring that if his cause was just, and he was thereby thrice armed, he would be invincible to any force of numbers or talent that should oppose him.

Unlike Hamilton in many respects, he was, as Jefferson said of Hamilton, a host in himself, and with a party or without, he was formidable, and it required the utmost circumspection of his opponents to guard the vulnerable points of their theory and practice against his masterly assaults. Whatever might be thought or said of his politics, the purity and fervency of his patriotism was never questioned; he proved his love of country by devoting nearly the entire period of his maturity and unsurpassed ability to her cause and her welfare. Like a pure fountain of sweet waters, his love of country, not in shallow, transparent, flimsy mouthings, was generous, and all men, whatever their persuasion or policy, could imbibe freely at this unfailing and refreshing spring, with profit to themselves and advantage to the State. Among the many lessons learned here from Nature's book he learned to be a disciple of Izaak Walton, — an exciting, healthful, and rational recreation, under the proper restrictions of reasonable times and seasons, — and Daniel Webster the angler, if not as famous and historical, is as familiar and interesting as Daniel Webster the foremost statesman of the age. It was in the midst of one of these favorite expeditions that he wrote the celebrated Hulsemann Letter, which may be regarded as the keystone in the triumphal arch of his diplomatic fame.

His lofty genius towered up like Kearsarge, under whose shadows he sported in childhood, conspicuous among his fellows, and, like this firm, immovable mountain, his head was bared to the pitiless, unreasoning, and incongruous elements; and, although Daniel Webster the mortal has passed away, he "still lives" in the brightest pages of his country's history, and in the affection of an enlightened people, and Kearsarge shall, by some fearful convulsion, fall sooner than the memory and fame of Daniel Webster!

He died Oct. 22d, 1852; and, as ten thousand bells pealed solemnly the knell of his departure, a nation bowed in grief at the portals of his tomb. Venerated he still lives in the example he has furnished, and is commemorated by the proudest monument yet reared to mortal, — the high esteem and deep affection of an intelligent and enlightened people.

"After a warm contest the Federal tickets for electors of President and members of Congress were elected by an average majority of not far from fifteen hundred votes.

"Daniel Webster was one of the congressional delegation at this time chosen; a man who, though then young, soon ranked among the ablest opponents of the administration and the war, and gained that high reputation as a cool, powerful, and eloquent debater which he has maintained during a service of twenty-five years, in one branch or other of the American Congress. Thirty years before this time he was born by the side of the Merrimack, — the son of a farmer. At school and in college he sometimes composed poetry, and displayed in his prose compositions a gorgeous fancy; but his first efforts at the bar were marked by a close, vigorous, and mature style, which indicated a preponderance of the reasoning powers over the imagination, and determined his character as a powerful logician, kindling but occasionally with the fires of imagination. He rose with a rapid flight, dazzling and astonishing, convincing and conquering.

"The bar acknowledged him as its head; the rival leaders of his own party made way for him in the race for distinction, and he was ushered forward at once to the first stations of responsibility and honor which they had in their power to bestow. Most of his mature years have been passed in the halls of legislation. He has discussed, either for the purpose of opposition or support, most of the important measures of government; and though his views have often failed to go in the popular support, and the correctness of them has been questioned by the ablest minds in the nation, yet he has left impressed on the memory of many, and stamped upon the records of public affairs, so many of those touches of genius, which in an age of intelligence will be preserved from oblivion, that the name of Webster, though he be consigned to the grave, cannot fall into forgetfulness." *

Kearsarge Mountain, a corner-bound of this town as well as

* Barstow.

many others by which it is surrounded, stands prominent. As if too great a prize for one alone, it is divided among the following towns, namely: Andover, Boscawen, New London, Salisbury, Sutton, Warner, and Wilmot.

The geological composition of Kearsarge is said to be principally mica slate, and is scarred and seamed by deep furrows, and much striated, as geologists affirm, by the drift, gneiss, and mica slate, combined with talc or isinglass, and limestone. Pyrope, regarded as one of the varieties of garnets, is found in the gneiss; and in Warner is a valuable talcose, or soapstone, quarry. Not only Kearsarge Mountain, but, indeed, all the other mountains and hills, and numberless other localities throughout the State, are rich in the variety and value of mineral products.

Whether the dream of the earliest settlers will ever be realized in the discovery of a New Hampshire El Dorado or not, time and the development of the, as yet, undiscovered and unknown resources of the State alone can determine; and it is believed that a judicious appropriation of a small amount for this purpose would in nowise affect injuriously the prosperity of the State, and might result in untold benefit. While the plough is the "mud-sill" of the superstructure of prosperity and wealth, the pick, smelting-furnace, and crucible are joists which give the edifice convenience, strength, and beauty.

"Kearsarge is *sui generis*. We never lost a good chance to climb a mountain, but never saw one like this. Its rounded top, embracing hundreds of acres, is a polished, undulating ledge. Some of these undulations are abrupt, forming, as it were, separate peaks, and, if the reader will pardon a slight exaggeration for the sake of a general truth, we may say there is hardly vegetation enough to support a squirrel, or loose rocks enough to drive him away. There are said to be indubitable evidences, based on facts and analogies, that the formation of this mountain-top is volcanic, — that great waves of lava have been thrown out at successive periods to cool in strata, and that this accounts for the wonderful regularity of the formation. There are also many well-defined depressions in the ledge, all running in the same direction, which geologists affirm were produced by drifting glaciers from the Arctic regions, bearing boulders, perhaps, in their icy grasp, in the time long agone when all this region was the bottom of the sea. The crater of the old volcano is pointed out.

"Right here in the middle of the State, overlooking one-half of its territory and three-fourths of its people, with nothing to obstruct the vision, 'swinging round the circle,' stands Kearsarge. The view lacks the grandeur of some others, but the world can scarcely produce one more beautiful. With a favorable atmosphere, nearly all the mountains in this State and Vermont may be distinctly seen. As it was, there were Ossipee, Chocorua, Carrigan, Whiteface, Lafayette, Moosilauke, Cardigan, Ascutney, the Green Mountain range, Monadnock, Uncanoonucs, and Wilson's Hill. Of lakes, we have Winnipesaukee, Newfound, Sunapee, and about thirty large ponds in the neighboring towns. Of villages, we have three or four in Andover, two in Wilmot, as many in New London, Sutton, Warner, and Salisbury, one in Henniker, Webster, and Boscawen, and two or three in Canterbury, Sanbornton, etc. The agricultural aspect is charming. We did not know our State was so attractive. Andover was too near to seem quite a prairie, but Wilmot, New London, and Sutton, on the north and west, and the other towns we have named, on the south and east, appeared nearly level; and the traveller will wonder where the hills are that he encountered in going to Kearsarge. Each public and private building in that populous section of our State stands out in bold relief; and the whole region, with its asperities softened by distance, seems a succession of beautiful and productive farms." *

The seasons as they roll afford, in turn, peculiar charms. Winter, with its wild and threatening aspect of clouds and frosty atmosphere, so sharp and penetrating that the trees and rocks are rent with terrific sound: under the shelter and protection of a hospitable roof, surrounded by congenial friends, and a generous warmth, it gives a sense of added comfort to scratch the thickly frosted panes and look out upon the fierce, wild storm; to hear the winds roar dismally and cheerlessly; to see the snow whirled by a frenzy of fury over all the land; to look across the broad and dreary waste of desolation, and feel secure.

In summer, when the concentrated heat wilts down the robust

---

* J. M. Campbell.

frame; when the very sweat of nature is evaporated from the oozy pores, and leaves her products seasoned; when the loud thunders, peal on peal, crash heavily, and bellow along the cliffs, and lightning darts in fiery forks and chains along the sky, and clouds roll in heavy, dark masses, like trooping fiends, so angry, fierce, and threatening; or the bright sun, rolling in unclouded splendor, — a scene like this is beautifully grand, but nothing can excel the gorgeous loveliness of an autumnal landscape view.

From the elevated summit of Kearsarge, when those two matchless artists, Autumn and Frost, have passed that way, and touched the foliage with their magic brush and incomparably magnificent coloring, the view cannot be excelled throughout the universe. Nothing intervenes on either side to obstruct the range of vision to the utmost limit of its power. Mountain, valley, hill, and dale, all robed in scarlet, purple, and gold, while a flood of mellow sunlight, pouring upon the scene, illuminates the forests with all the hues and shades of the rainbow, delighting the eye — never satiated or tiring — with the magnificent spectacle of a more than fairy land.

Creeping unobtrusively and unobserved through the valleys, up along each acclivity, through the great forests, and over the hilltops, the unequalled painters have touched the brush to the trailing vine as well as shrub and bush and giant trees; nothing escapes their attention, and now, where everything was but yesterday so uniformly green, crimson and gold and all the intervening tints adorn the landscape. Looking to the eastward, this brilliant scene of beauty is diversified and enhanced by the broad Merrimack meandering gently through the wide green valley far below, like a glistening baldric, while nature everywhere seems dressed in her liveliest and loveliest habiliments, and in the midst of all this lavish display of splendid variegation of adornment stands the unchanging perennial, superbly ornamental in its sombre panoply by contrast with its gaudy surroundings. Sublime and enchanting as an autumnal scene like this cannot fail to be, it is strange so many greet this season sorrowfully, indulging in mitigated lamentations, and discoursing solemnly of the "sear and yellow leaf" and "gloomy days" which "chill the spirits and turn the heart to loneliness and sadness." Notwithstanding poets, orators, and scholars almost invariably affect this emotion, there are those who fail to receive similar impressions, but, on the contrary, witness the advent of this season with the liveliest emotions of joy and pleasure.

As the time approaches for the termination of the brief period of nature's life, and the whole earth seems an interminable paradise of beauty in its gaudy attire, it not inaptly symbolizes the transition of the genuine Christian, the good man, the true man of God. When his time of dissolution comes, he is, instead of being overwhelmed with sorrow, fear, and remorse, cheerful, buoyant, and happy. He grieves not to depart. The good deeds of his life illuminate the dark valley, and shed a halo around him, which, like autumn leaves, robes him in brighter raiment as he falls into the tomb.

As if rejoicing at the maturity and perfection of its beneficence, having budded, bloomed, and ripened to the glory of God and the good of mankind, the grand old tree appears in its most attractive attire at the very moment of dissolution. So it is with the spirit of the just; glowing with sublime effulgence in the consciousness of duty done, and faith in the promise, and grateful for the prospect of immediate and full fruition, the beautiful panoply of joy and peace and righteousness adorns the closing scene. Nor is this the complete similitude. The boundless wealth of coloring, which is one of the crowning glories of the life and death of nature, points, unerringly, to another life, a renewed existence, an opening springtime, where they shall bud and bloom again in pristine beauty and glory.

Thus the tried and true spirit, undoubting and undismayed, views, with the ecstasy of a hitherto unfamiliar joy, the sign of the reopening of an unfading spring of perennial bloom, when rebudding on the Tree of Everlasting Life, nevermore to know the "sear and yellow leaf," or the "gloomy days," unchanging in the pure raiment of unfading glory, proclaims through an eternity of bliss the matchless charm of autumn leaves!

Boscawen lies on the west side of the Merrimack River, which is its eastern boundary. It is a very handsome and interesting town, its principal village being a broad avenue, about two miles in extent, ornamented with fine old trees, which give a picturesque and pleasant air to the place, and afford a welcome shade in the heat of the summer months. The Blackwater, a tributary of the Contoocook, has its course entirely through this town, parallel with, and about four miles from the Merrimack, and, having falls, it affords many mill privileges, which are well improved. The original settlers erected a log fort, one hundred feet square and ten or twelve feet high, to protect themselves against the savages, and for more than twenty years this afforded them their only protection.

In 1746, Thomas Cook, a negro, was killed by the Indians, and a man named Jones was carried to Canada, and there died. In the year 1754, Nathaniel Maloon, and his entire family, consisting of his wife and five children, were taken captives, and carried to Canada, where they were held in servitude nearly four years. In 1756, Edward Emery and Ezekiel Flanders were killed while on a hunting expedition, and there was no safety or security anywhere except under the immediate protection of an armed guard.

The first bridge over the Merrimack, below the forks, in Franklin, spans the river on the highway from Boscawen to Canterbury. At the confluence of the Contoocook with the Merrimack, which occurs near the southern boundary of the town, there is an island belonging to Boscawen which will remain forever famous for the heroic daring of a brave and determined woman. The terrible tragedy which was enacted on this island one hundred and seventy years ago, is unparalleled by the creation of the most fertile imagination, and will live in the annals of New England as long as heroism shall continue to be a virtue.

On the fifteenth of March, 1698, Mrs. Dustin, her infant, a week old, and Mary Neff, her nurse, were taken captive by a party of Indians, at Haverhill, Mass. Mrs. Dustin was driven from her bed, and compelled to accompany her savage captors. After they had proceeded a short distance, the child becoming a worthless burden to them. one of the number put a period to the troubles and sufferings of the innocent by dashing its brains out against a convenient tree. Proceeding up the Merrimack with the nurse and the mother, sick, exhausted, and overwhelmed with anguish at the sight of this shocking cruelty, the party encamped on this island, which is known as Dustin Island. The landing here was designed only as a temporary rest for the savages, who intended to continue their journey up the river to an Indian settlement, where they proposed, for the amusement of themselves and confusion of the prisoners, that they should perform the ceremony of running the gauntlet.

Aware of the cruelties practised upon their victims, Mrs. Dustin, in her desperation, formed a determination to exterminate, if possible, the savages, and liberate herself and fellow-prisoners. Her companions, the nurse, and a lad, Samuel Leonardson, who had been previously captured by the same party near Worcester, were informed of her intentions, and instructed, as well as possible, in the details of the enterprise. The Indians, worn down with fatigue by their long and rapid retreat up the river, after refreshing themselves, sank on the ground, and soon fell into a profound slumber. Now was the favorable opportunity, and the heroic woman, undaunted by the strength, skill, agility, and numerical superiority of her captors, proceeded to execute her daring scheme. By the aid of the nurse and lad, and the murderous implements of the savages, she despatched ten of them. A woman of the party escaped, and a boy was purposely left unharmed. Scalping the slain, and freighting one of their best canoes with a few supplies, arms, etc., this weak but resolute party headed their frail bark down the Merrimack towards home, and arrived safely in Haverhill to the infinite surprise and indescribable joy of their friends.

The self-sacrificing devotion of Mary Neff, in her determination to remain with her mistress and share her dangers, rather than to escape, which she could have easily done, thus leaving Mrs. Dustin alone to her fate, appears not only to have been providential, — as the terrible tragedy could not have been successfully enacted without her assistance, — but is worthy of the highest commendation.

In the year 1828 a portion of each of the towns of Andover, Salisbury, Northfield, and Sanbornton was taken and erected into a separate town called Franklin, which has, during its comparatively brief existence, exhibited considerable enterprise, and is a flourishing and progressive community.

For several years past, as well as at present, Franklin has advanced in population and resources. A variety of mechanical works, hosiery-mills, etc., are in operation. Aiken's famous knitting-machine was invented and manufactured here, and has been a source of wealth to the inventor, as well as a great labor-saving machine and improvement in the manufacture of knit over and under garments.

Franklin is the terminus of the Bristol Branch Railroad. The Northern Railroad — Concord to White River Junction — passes through the town, and, as the road is above the roofs of many of the houses, the attention of the passenger is arrested in passing through it, and in those who are constitutionally timid a slight nervousness is liable to be superinduced.

Arriving at the forks of the Merrimack, the topography and condition of the country present so marked a difference from that at its source as to at once arrest attention. Broad intervals and gently undulating uplands in a high state of cultivation, in richest verdure clad, betoken an excellent soil and a thrifty and industrious farming community. Thus pleasantly located is the enterprising village of Franklin, about one-fourth of a mile from the forks of the Merrimack.

Here at the forks the bony shad and the sinewy salmon, in the palmy days of the migratory tribes, parted company; the former taking the dark, tepid waters of the lake, the latter the cool, bright current of the Pemigewasset, whose countless cascades and cataracts, and rumbling, tumbling, foaming falls afforded this unequalled fish the element he preferred, and the exercise in which he so much delighted. Pitting his great strength and agility against the resisting power of the rapid current, he headed towards its source among the mountains, where, like other sensible beings, in the secluded retreat of the pure, cold pools, darkened by the shadows of the great surrounding mountains and forests, he passed the dog-star period in security, comfort, and repose. Shooting with the velocity of an arrow across the boiling, turbulent waters of the falls, he regained his summer residence from year to year, until shut out from the land or water of his nativity by the impassable barrier at Lawrence.

CHAPTER IV.

The Forks. — Winnipesaukee Lake and River. — Pickerel Fishing. — The Wiers. — Laconia. — Capt. Lovewell. — Centre Harbor. — Moultonborough. — Red Hill. — Sandwich.

HAVING arrived at the forks of the Merrimack in Franklin, namely, the Pemigewasset and the Winnipesaukee, it will be proper to turn attention to the latter stream and its great reservoir. It is at the confluence of these two streams that the Merrimack takes its name, — a name, whenever and wherever heard, suggestive of bright waters, industrious, prosperous, and happy communities, great factories filled with intricate and delicate machinery, and bright eyes and nimble and cunning hands to guide its movements.

Lake Winnipesaukee, the source of the river of the same name, is a large and magnificent sheet of water. It may be regarded as the grand central plateau of New Hampshire waters. The name is of Indian derivation, and has not unfrequently been erroneously interpreted as signifying "The Smile of the Great Spirit." This, however, is incorrect; it being rendered literally signifies the "beautiful water of the high place." The lake is something more than twenty miles in length, and affords a scene of unsurpassed loveliness. Like the glimmer glass of Cooper it spreads away like a liquid sheet of burnished silver in the bright sunlight, and forests rising high along its emerald banks mirror and reproduce themselves deep in its placid bosom. A calm serenity sits enthroned upon its polished surface, except when moderate breezes stir it into dancing ripples, or strong gales move it to gentle undulations.

This lake has many notable features. It is reputed to contain, like other famous waters, the inevitable three hundred and sixty-five islands. Whether this is so or not, the islands are numerous. On some of them are many farms of excellent and unusually productive land; others are used solely for pasturing, and herds of cattle and large numbers of sheep here find a splendid summer resort, securely enclosed, no trouble to the owner or his neighbor, and at the same time entirely self-sustaining; others are used solely for the gathering of picnic and excursion parties which resort here from long distances as well as the immediate neighborhood of the lake; and public-spirited or speculating individuals have erected permanent buildings, furnished with all the modern conveniences for innocent recreation for old and young, for religious societies as well as for the world's people; and here, throughout the summer season, crowds disembark from the steamers daily almost, having fled from the sweltering brick walls of pent-up cities, armed with all the needed supplies and luxuries to enjoy a holiday of social and rational pleasure, to indulge in the exciting and exhilarating sport of fishing and other amusements, and to drink of the waters of the translucent fountains, cool and refreshing from the mountains.

There are, also, many islands in this magnificent lake, luxuriant in majestic forest-trees, wild in matted and tangled undergrowth, like monster emeralds in a silver setting, or like an Oriental picture, the permanent haunt of the rattlesnake, and the temporary resort of the strong-lunged loon and other aquatic fowl. Considering the great size of this lake, its water-shed is extremely limited, so much so, that it is a great wonder how it maintains as steadily as it does its maximum depth. The topography of the surrounding territory would appear to the view well adapted to supply the feeders of this great body of water, but no stream of any magnitude finds its way into it. A narrow strip of territory skirting the lake shore supplies the usual small brooks, and these comprise the sum-total of its affluents. The extensive country stretching away to the north some twenty or thirty miles to the Sandwich Mountains, whose inclination is to the southward, including the country some distance to the east and west, would seem to be the natural supply for the lake; but this is not so. The Bearcamp River rises among these mountains, and, after approaching within a few miles of Winnipesaukee, trends eastward and discharges its waters into the Ossipee Lake, from thence disgorged through the Ossipee River into the Saco.

It will be seen that, while the Ossipee, an inconsiderable body of water, only six or eight miles in its greatest length, receives the waters of two rivers, — the Bearcamp and Pine, — the Winnipesaukee, more than four times its size, is fed by no stream of any great consequence, the outlet of Goose or Measley Pond, which falls into the Winnipesaukee at Meredith Village, being one of its largest affluents. Discharging an important river, and constantly maintaining full banks, its poverty of resource, and the extensive surface presented to the depleting process of evaporation, this lake is a problem, the solution of which can only be reached by the conclusion that it is fed by invisible subterranean springs, very large and numerous.

Nearly all of the territory on the eastern side is drained by the Cochecho,. which has its source or sources in the immediate vicinity, and it is well established that the lake can easily be turned and drained into the Cochecho instead of its present channel, thus making a vast difference in the power of the noble Merrimack; for, without this important tributary, and more especially without this indispensable reservoir, the supply would fall short. Experience and a wise forecast combined have led the great companies along the river to provide artificial means to reduce the whole extent of this vast reservoir several feet, to meet the contingencies of hot weather and dry seasons.

The lake abounds in fish of many varieties; the lake trout and the pickerel being the most important and valuable. Cusk are much prized by many, and perch and horned-pout may be taken at any time, by the most inexperienced anglers, in quantities to suit. A variety of salt-water fish were some years since placed in this lake by experimenting parties; but as nothing was seen or known of them afterwards, it was presumed they found their level either in a "watery grave" or through the channel of the Winnipesaukee and the Merrimack in a more congenial element, the briny deep.

Lake trout are taken, but not so plentiful or so large as formerly; still an occasional laker of gratifying dimensions is hooked, varying in size from three to thirty pounds. In June, 1868, one was taken out weighing seventeen and a half pounds. The sportsman is always liable to procure a coveted contest with at least one of these stout and powerful fellows. Deep fishing is the only mode of taking them; and, proceeding to the approved fishing-grounds in a skiff, with an oarsman to assist, the sportsman takes the soundings, which vary from sixty to eighty feet, then lowers the hook, attached to a strong line, sufficiently near the bottom, and nervously awaits the issue. Unmistakable demonstrations from the deep, warning him that the prize is hooked, the sportsman now turns his attention and skill to the work of securing the trout. Practice and prudence are now required for a safe and successful issue of the contest. The fish, discovering his condition, makes tremendous efforts to free himself, and requires to be reeled in and run out with a firm hand and a taut line until exhausted, or, as it is termed, "drowned," before he can safely be secured.

But the pickerel is the most plentiful and the most valuable of the lake fisheries. Summer and winter alike the business of catching pickerel is prosecuted with great success. In winter, it is estimated that five thousand lines are the average of the daily set, whenever the weather will permit. Ice fishing may be described thus: —

A party of two proceed with axe and chisel and cut from thirty to fifty holes in the ice, this being the number they can properly attend to. They then proceed to sound the depth of water in each hole successively, and set the hook. The depth of water is usually from three to five yards, and the line is gauged a little short of the depth required; a movable signal of colored cloth is attached to the line; a live, red-finned shiner hooked directly under the dorsal fin and lowered through the hole; the other end of the line is secured firmly to some object, and the slack of the line near the signal is attached to a slender twig a foot or more above the ice. All the lines being set, the fishermen carefully watch the going down of the signals, which occur when the pickerel takes the bait. In this manner they often secure large fares, and have lively work to care for all the lines properly. Many of the inhabitants about the lake have shanties constructed on a sort of sled, provided with comforts and accommodations for sleeping and cooking, and, when the ice is sufficiently strong, oxen and horses are attached, and they are hauled upon the lake to the fishing-grounds, and rented to parties, affording a considerable revenue to the enterprising proprietors.

Summer fishing is, however, if not more profitable, much more comfortable and exciting. At this season the numerous and extensive estuaries and coves which indent the shore are all alive with pickerel, prowling in quest of prey among the reeds and rushes along the marshes, or watching for the little blue fly securely hidden under the lily-pads, *nuphar advena*, that cover the surface with their broad leaves to the very brink of deep water. Seated in the bow of a boat, with a careful sculler in the stern, armed with a twenty-two-foot rod and line to match, the boat pushes carefully along fifteen or twenty yards outside the line of water-grass; the hook, baited with a frog's leg peeled, is handled out and made to ricochet along the surface. The pickerel whirls out from his hiding-place, and, with a powerful muscular movement, rushes upon the prey; at this time, apparently, in a completely inverted attitude. Experts differ as to the best mode of proceeding from this point, — some contending that, as he instinctively strikes his jaws together with great force to despatch his prey before swallowing it, it is impossible to pull the hook away from him : consequently he is sure to be fastened. Others endorse the theory that he never swallows the bait until satisfied that it is both killed and palatable. Convinced that the frog's leg possesses the latter quality, he is allowed to proceed with it in his own way. Having got it in his mouth, he invariably retires to the vicinity of the bottom, but a short distance, to test its quality. It is necessary to keep a tolerably straight line on him during this time, as he always moves the moment he swallows the bait, and this movement is a signal to take him in. It is believed that the latter mode is much the surest and safest, as it hooks him stronger and more certain, although practice is required to land him safely in the boat. The pickerel is an excellent pan-fish, second only in value and quality to the trout, and is, like the latter, a most important item of New Hampshire inland fishing; and, although he is not found in brooks, or in the cold waters of the mountains, like the trout, he is indigenous to all the other waters of the State, rivers as well as lakes and ponds.

One feature, not, however, peculiar to these waters, but which adorns and beautifies them, is the starry magnificence of the milky way of the great white water-lily of midsummer, which here flourishes in more than Oriental luxuriance and splendor, and may properly be termed the queen of the New England flora. Gathered in huge bunches, with long and leathery stems like coils of lacings, and placed in capacious vases on parlor tables, they are regally grand to look upon, and emit an incomparable delicacy and richness of fragrance, an intoxicating perfume pervading the house; but here is not the theatre of their greatest attraction. Floating buoyant and graceful along the lake, they seem an endless city, a great London, in miniature palaces of ivory and gold.

The Victoria Regia may be a superbly royal flower, the Egyptian Lotus may be famed above all ordinary productions, but the waterlily of New England can never be surpassed in grace and beauty, in the untold wealth of its unequalled fragrance, or in the hearts of all classes or sorts of people. The little barefoot hails them with no less eagerness and delight than the lady of refined taste, or even the grave and grizzled merchant prince. Shutting the gates to their golden citadel at close of day, as if to avoid the darkness and gloom and the contaminating miasma of night, and opening their bright eyes with the opening day, they typify the pure and noble in human life and character, and symbolize by their natural beauty and purity, the genuine Christian heart, bestowing on all alike, the humble and the high, their generous fragrance.

Several steamboats ply across the lake, bearing immense numbers of tourists and pleasure-seekers towards their destination,—the mountain region,—and there can be no more perfect paradise on earth imagined than the hurricane-deck of one of these floating palaces on a hot summer day, swiftly gliding over the lake; to see the gems of islands dotting all the vast expanse of water, and the white water-lily gently floating, securely anchored by its stem, or humbly dip its head beneath the resistless boat, then, rising, kindly waft its grateful fragrance on the wings of gentle zephyrs over all the scene, while the bright waters, the deep-green foliage of the shore, the shrill scream of the loon in the distance, coming like a thrice-told echo to the ear, the mountains rising in hazy grandeur, scarred with ghastly rifts and seams, verdant meadows, cornfields, and green pastures of flocks and herds,—all these combined, make up a landscape view, a gorgeous panorama, which may be seen and enjoyed, but not described.

The waters of this lake, although clear, pure, sparkling and sweet, differ in color from the Pemigewasset or Merrimack, being darker; the difference is so marked as to be readily told fifty miles below where the great reservoir is tapped. New Hampshire is justly proud of her inland waters, foremost of which are the Merrimack and Winnipesaukee. In this feeling all the people of every section cordially unite. Nor is she selfish in this matter; for she pours out her boundless wealth of waters into the lap of her sister State, the latter equally with herself enjoying the benefits of the unrivalled Merrimack and all its tributaries. They are not only of great importance and value at the present time, but are of great historic interest. The extent and variety of manufacturing and mechanical business carried on by the indispensable aid of these waters is one of the principal sources of her thrift, and a main avenue to that position of independence and wealth she now occupies and enjoys, in spite of the adverse circumstances of high latitude and ungenerous soil. Some of her noblest sons, best known to fame, were raised along these waters, and some have found in this beloved valley a final earthly resting-place; others, less known to fame, yet no less worthy men, who, although perhaps untutored, entertaining a just appreciation of liberty and independence, marched bravely through from Bunker Hill to Yorktown, making a more telling and conspicuous mark upon the enemy than on the muster-roll.

It is no new thing that the lake should be the centre of attraction, or a thoroughfare for a great number of people. It has been so for hundreds of years, the only change being in the character and object of those, who, in the rolling decades of time, have made their way hither. Indians came here to collect and carry away their subsistence, provided by the beneficent hand of nature; then followed next his relentless and exterminating pursuer, and lo! the poor Indian was faced westward to the land of sunset, where the sun of his race will presently set in darkness and in blood, to rise no more. Then followed the woodman's axe, — that mighty leveller, — toil and mechanism, enterprise, trade and pleasure-seekers; and, finally, they came to take the very waters that bred the Indian's meat.

Winnipesaukee is an Indian name, and is derived from winne (beautiful), nipe (water), and kees (high), and auke (a place); and the Indian tribe that inhabited the territory adjacent to the lake took the name, the Winnipesaukee, as well as the lake itself, — "Beautiful water of the high place." Nothing can exceed the native simplicity of this descriptive title given by these untutored savages, showing an appreciative and admirable taste for the grand and beautiful, which may profitably be studied by some enlightened people whose nobler natures seem to be swallowed up in the pursuit of lucre, and who see no beauty in anything, unless it has a visible dollar in it. Another interpretation has frequently been given to this word, which, entirely inaccurate, is yet exquisitely beautiful, and has, no doubt, been suggested by its eminent appropriateness: "The smile of the Great Spirit." Emphatically a "smile," equally to the Indian and the white man, of the Great Spirit, this lovely lake has been. To the white man it has been a teeming and inexhaustible source of pleasure and profit, and is literally a fountain of untold wealth to him and his posterity. To the Indian it was a granary and a treasury, affording him the means of subsistence, material for all his implements, primitive, yet indispensable to him, whether on the warpath, the chase, or in the economy of his family and domestic arrangements. It also afforded him the means of travel and communication with his neighbors; and the alluvial bottoms, quite numerous about the lake, cultivated by the squaws, served for his scanty tillage, providing him with gourds, and also an opportunity to indulge in that indispensable Indian institution, the green-corn dance.

Like all the tribes of the Merrimack valley, the Winnipesaukees were tributaries and confederates of the powerful and warlike nation of the Pennacooks, and it is said that the established or customary holidays of the Indians were a sort of movable feasts of the Pennacook confederacy, alternating between the leading fishing-places, such as Pawtucket, "Namoskeag" and Winnipesaukee, and the prominent points where game was collected, and the agricultural districts, although the latter, being few and very limited, were of secondary and inconsiderable importance or consequence. The Winnipesaukees maintained a permanent fish-weir, or ahquedaukenash,* as they called it, at the outlet of the lake where all the tribes were cordially invited to assemble for the spring and fall catch, and all who could, attended. One remarkable fact connected with the fishing at this place was that nothing but shad were captured. Of all the migratory tribes that left the sea in the spring for their spawning-ground, the shad alone reached the lake, the eel seeking the congenial mud; the alewives, being small fry, took to the smaller brooks and ponds, where the absence of large fish rendered it less hazardous to their *ova*. Salmon and shad proceeded together, until, reaching the forks of the Merrimack in the town of Franklin, they parted company; the salmon heading up the Pemigewasset, which, having its sources high up among the mountains, and its course

---

* The last syllable of this word — *ash* — attached to an Indian noun, signifies the plural number; applied to inanimate things.

through a long stretch of shady forests, afforded those cold waters, ripples, plunging torrents, dark pools, and wild whirling eddies, in which this magnificent fish so much delighted, and where they found their favorite spawning-grounds. Shad, as peculiar and unanimous in their tastes, preferring warmer and more quiet waters, took to the Winnipesaukee, and through that river passed into the lake in countless myriads, where there was ample room and favorable opportunities for the development of the millions of eggs that were required to supply the waste of the original stock, constantly depleted by ravenous fish, and a still more insatiate enemy, the red-skin.

"Ahquedaukenash" signifies, literally, a dam, or stopping-place, and was constructed in this wise: Large granite boulders were placed in an irregular line across the river, the boulders representing the angles of a crooked rail-fence, and at a proper distance below the falls. Wherever it was practicable, strong sapling stakes were driven into the bed of the river, and used for the same purpose, or took the place of rocks: but at the outlet of the Winnipesaukee this was impracticable, owing to the solid character of the river-bed. Having thus prepared the foundation, the rocks being some ten or twelve feet apart, a netting was then woven of twigs and tough and pliable bark, with meshes sufficiently close to prevent the fish escaping. This was strung entirely across the river, above and against the rocks, excepting a space between one or two of the rocks or stakes, these being left open for the fish to pass through in their progress up the river; through these openings the whole force of the fish must and did pass.

As few of them scaled the falls until after repeated efforts, and the rapidly advancing "school" crowded steadily through the opening, it follows that the pen, or ahquedaukenash, was soon full. Now was the time for the Indian shad-catcher. Expert fishermen, and such others as were selected and appointed for the purpose, manned the canoes and pushed boldly out among the pent-up prisoners, and with spear and dip-net lost no time in filling the canoe, in this regard illustrating the old maxim, "Make hay while the sun shines." Returning freighted heavily, they handed them to the squaws, who stood ready, knife in hand, to split the fish, and hang them up to smoke for winter on the centre-pole of the wigwam, or laid them out to dry in the sun on improvised flakes.

Sports were indulged in, varying nearly from the sublime quite to the ridiculous, serious, laughable, and grotesque: speech-making, feasting, and frolicking. Generally the soft and tender emotions were indulged in, and the young lover poured the tale of his long-cherished affection in a deep, overcoming guttural into the willing ears of his tan-colored charmer, and bent down a strong sapling to symbolize the tenacity of his affection and how much it could yield without giving way; or, perhaps, to insinuate how he would "double up" any luckless wight who should have the temerity to offer to put a ring in her nose.

Holidays and holiday-rites were frequent among the Indians, — the male portion of them, — the squaw's first genuine holiday coming generally when she died. While living she was a mere drudge; her sphere was to do all the work of every description in doors and out. She was first to recognize the unquestioned superiority of the status of her lord and master, anticipate and provide for all his wants, keep the wigwam in (Indian) shape, prepare the game and fish for preservation, or for present use, care for the pappooses, cultivate the gourd, the pompion, and the maize, and grind the latter for table use, and then devote herself assiduously to ornamental needlework. In this they were tasty, skilful, and active. Beside moccasins, belts, and head-gear, they wrought broad blankets with beads, fancy feathers, and bones of animals in fanciful and fantastic patterns, in many cases showing a genius for combination and display artistic and meritorious. It may seem strange, with all their other duties, arduous and severe, toilsome and wearing, and in the face of the discouraging fact that none of this finery was for their own adornment, but for their imperious and idle lords, that they should have found either time or inclination to prosecute the ornamental art to the extent they did; but it must be remembered that tradition and custom made her uncomplainingly and willingly subservient. Even this is less than the fact: for she taxed her ingenuity and strength, but never her patience, to devise or invent some new surprise, some ornament more beautiful and rare, with which to greet him on his return from the war-path or the chase. That the Indian woman should differ so widely from the white in this respect, in her sheer unselfishness in abdicating regard for personal adornment in favor of her husband, in her self-sacrificing devotion, — all this may not be strange, for was she not a barbarian and a pagan?

The tribes of the Merrimack valley, consisting of the Winnipesaukees, Pennacooks, Souhegans, Nashuas, Wamesits, and Agawams, of which the Pennacooks were the acknowledged head, were united in a powerful offensive and defensive confederacy, to which belonged, in addition to those named, other Indians, both in New Hampshire and Maine. All of these were undoubtedly entitled to indulge in fishing, under Indian regulations, in any or all the waters of the territory of the Pennacook nation. With hooks made of bone, dip-nets, spears, and other rude devices, it was prosecuted from necessity, but generally on a small scale and with indifferent success, except at the ahquedaukenash, where the unfailing supply and the assembling of the representatives of all the tribes made this a festival of great note and consequence among the Indians.

The permanent ahquedauken at the foot of Lake Winnipesaukee was of substantial construction, portions of it remaining long after most of the emblems and monuments of aboriginal occupation had gone to that oblivion to which the doomed projectors were ultimately consigned; but though the Indian is gone, and the "stopping-place" is gone, and the salmon and shad are gone, the name still remains, and the fish-weir of the native, far away in the heart of an unknown wilderness, is now familiar throughout the land as "The Weirs," a crowded thoroughfare on the route of travel to the mountains, at the foot of the lake. How changed the scene! Where the red man grubbed for a few stalks of stinted corn, the ploughshare, bright with use, turns up the teeming soil; the rude fish-weir has given way to skilful artificial arrangements for taking the water from the lake at will; the frail canoe has given place to the palatial steamboat, and the primitive and picturesque costume of a bearskin tied about the waist became unbecoming and disappeared before the resistless march of fashion; and elegant fabrics from the Eastern World, with length and breadth enough to gratify the most ardent admirer of extravagant proportions, mark the difference between the economical and scanty provisions of mother nature and the productions of art and skill. In brief, the change of proprietors has wrought a change in all else, and it may be truly said that countless ages rolled away, and left it still a howling wilderness, while the lapse of two hundred years under the beneficent influence of civilization has made it blossom like the rose. Where the fearful war-whoop resounded from crag to cliff, waking the echoes of the dismal forest, and the prowling wolf howled a dire homesickness to the benighted and solitary forest ranger, and the splash of the paddle as the dexterous red man "feathered his oar," or the plunge of the gigantic moose as he laved his unwieldy bulk in the tranquil waters of the lake, or the loud scream of the loon as it called to its mate across the tranquil bosom of its waters, — where sounds like these, and only such as these, were heard, the iron horse now tears along, bearing his endless trains of humanity and merchandise, the pleasant sounds of spindle, loom, and anvil are heard through all the day; and on the very spot where stood the red man's wigwam — a rude, uncomfortable hovel — are flourishing villages of comfortable and elegant dwellings, fine stores, large and convenient workshops, and an intelligent, industrious, enterprising, and thrifty population. What has wrought this great change? Precisely the same generous earth exists that grew the Indian's scanty vegetable products; the same pellucid waters spread out like a broad mirror and roll their never-ending tide down to the unsated ocean; the same mountains, towering high, moody, and silent, stand guard and sentinel around: the same translucent atmosphere pervades all things and places, and above all the same genial sun rolls on, as it has for numberless centuries, warming and fructifying; and the seasons come in turn as they have always done. Then why this change? Has there been a new dispensation of providential favors? No; it is the superior organization, mental and physical, of the Caucasian race. Perseverance, native skill, untiring industry, handicraft, cunning, under intelligent direction, and, above all, the genius of the white man's government, — liberty restrained by law that it may not degenerate into lawlessness and license, — these are the predominating and pre-eminent causes of the immeasurable difference between the red man and the white.

On leaving the lake, Lake Village, an enterprising and flourishing village in the town of Guilford, is reached at a short distance from the foot of the lake. Possibly this observation may be incorrect; for, although the outlet of the lake proper is a river most indisputably, having all the characteristics, a rapid and unchanging current and a pebbly bottom, confined to a channel, and having, also, the legal status of a river, yet a little further down it debouches into a broad lake, and may not, perhaps, improperly be regarded as an extension of the lake proper, especially as it maintains the character of a lake for ten or fifteen miles below the above-named village.

When the town of Meredith was divided, the southerly part, including what was formerly known as Meredith Bridge, was set off and incorporated as a separate town, under the name of Laconia. This is a large and flourishing town. About a mile below Lake Village is the principal village, on the opposite side of the river; and, like Amesbury and Salisbury, in Massachusetts, it is difficult to decide where one terminates and the other begins. Laconia was long since the name by which all this region of country was known, and it is presumed that this fact, together with the other, namely, the town of Meredith being an unwieldy territory, was the reason why the town was divided. In this village, as in the adjoining one, there are extensive manufactories. The largest companies at the present time are, the Belknap, capital $100,000; Guilford Hosiery Co., $50,000; Ranlet Co., $15,000. In Lake Village are situated the repair-shops of the Boston, Concord, and Montreal Railroad; and, scattered through both, are mechanical works varying from shoe-pegs to locomotive engines.

Only on mature reflection can the incalculable benefit of these manufactories located along the route of this river and its tributaries be realized. Flourishing cities and beautiful villages have sprung up from the very rocks and sands which were hitherto valueless territory, and which, but for this river, might, perhaps, have remained forever an unproductive waste. It has collected together a vast capital, inventive genius and talent, and created an immense number of skilful artisans, mechanics and workmen, and a thrifty and industrious population, thus making the capital, skill, and labor reciprocally remunerative. These communities have afforded the neighboring country a profitable and steady market wherein to buy, sell, or exchange; made cotton and woollen fabrics so abundant as to be afforded, elegant and cheap, to distant and less favored people, and secured to the United States a well-merited reputation in this branch of business, equal in all respects — beauty, style, quality and finish, durability and cheapness — to that of older and more experienced countries. Indeed, the Merrimack River manufacturers enter confidently into the expositions of the whole world, and bear away in triumph the gold medal and other evidences of superiority over a prodigious array of reputation, capital, organization, and skill long established. Thus it is that the immense business created by this fine river is not only a benefit at home, but has its beneficent ramifications throughout every channel of the country.

How long the Merrimack has maintained its present condition and appearance it is impossible even to conjecture. Certain it is that at some period, far in the dim, distant past, the river was one continuous chain of lakes, whose barriers, being worn by water, ice, and drifting wood, have successively given way until this whole system of collected waters was drained and ultimately reduced to, and confined within, the present banks. Extensive alluvial deposits indicate the former character of these waters, and their location and dimensions can still be distinctly traced, while far below the surface are found well-defined vegetable deposits, logs, and other foreign matter brought here and left, perhaps, for evidence of these facts, far away from the present channel of the river. If more proof were needed, it is supplied by the peculiar stratification of the soil, which is regarded by scientific men and geologists as conclusive on this point.

The great bay, and even Lake Winnipesaukee itself, are cited as still existing portions of the former condition; and who shall say when the time will come that from the northern extremity of the lake to the forks at Franklin nothing but a rapid and noisy river shall mark the former existence of these invaluable and beautiful waters. It has already been observed that the neighborhood of Lake Winnipesaukee was a famous place for the Indians to congregate. From here they went down to Cocheco, Dunstable, or other places, to barter with the whites, or swept down upon the almost defenceless colonists in their most fearful raids, impelled by a savage barbarity and a thirst for human blood. At that time the whole of the vast region stretching away to the north, beyond the head-quarters of the Pennacooks, now Concord, was unexplored and entirely unknown, except the French settlements in Canada; consequently the settlers along the Cocheco, the Piscataqua, and the lower Merrimack were without even the ability to conjecture the hordes of barbarians this wilderness was capable of producing and pouring down at a moment when least expected upon the defenceless settlers. This want of knowledge of the number and condition of their adroit and implacable foes annoyed them exceedingly, and caused them to become vigilant, active, and always on the alert, and was the cause, even more than their numbers or courage, of constantly frustrating the exterminating designs indulged in by the savages.

Had the tribes proved as numerous, determined, and well prepared as was feared, the progress of the colonies, not only in New Hampshire but Massachusetts likewise, must have been greatly impeded, if not altogether stopped. As it was, they would often swoop down upon the scattered settlements as swift, sudden, and unexpected as the hawk, and if they often met his fate they still managed to inflict material injury upon the victims. While this state of things existed, men were cut down in the forest, and in the clearing. Everywhere they went armed, often throwing out pickets to protect them while at work. Families awoke at dead of night to hear the fearful war-whoop, to see their homes enveloped in flames, and the deadly tomahawk and the dreaded scalping-knife gleam in the light of the burning dwelling. Without succor, and with no hope of escape, to be resigned and to die was all that was left for them, unless it was desirable to have a little barbarous diversion in the shape of torture. In the event of torture being determined upon, savage ingenuity was taxed to devise or invent a refinement of cruelty which would afford frightful and ferocious amusement.

Midnight massacre, and the little less preferable captivity, were the ghosts of every household; and the man who had a family to protect and provide for, could relax from his duties only to turn his mind upon the terrible fate which possibly, nay, probably, awaited himself and his. But this state of things was not to last forever. It was only a question of time. The indomitable white man was not to be exterminated by a race so untutored, nor his spirit subdued, nor his enterprise 'long checked, nor his prospects for a glorious career blighted. Not he. Facing a severe climate, fierce wild beasts, an unsubdued wilderness, and ferocious and implacable human foes, he went steadily forward. Slow at first, to be sure, but still he went on, until, by his irresistible energy, he penetrated farther and farther into the depths of the unexplored wilds, driving his enemy before him, until at length his foot was so firmly planted that his desperate and terrible foe, giving up the contest in dismay and despair, retired beyond the reach of the invader, and left his birthright and his soil, in the undisputed possession of the paleface.

The wheel of fortune is ever rolling, and he who stands on its topmost segment to-day may experience a revolution so sudden and overwhelming, that he is crushed, as by the car of Juggernaut, into the dust to-morrow. So it was with the doomed race. With no idea of retributive justice, with no thought of the possibility of reverse, or a change of the situation, the Indians were remorseless, sparing neither age, sex, or condition; but the war-club, the knife, tomahawk, and arrow, the conflagration, torture, and the dreaded captivity, were meted out unsparingly whenever opportunity offered; but now the other side had the "innings," and laid themselves out for a large score rather than for fancy playing. Having been trained in his school to the perpetration of barbarities, cool and deliberate, the white man tried his hand at articles hardly less savage than their own. A price was set upon his toplock the same as on the wolf, and other wild animals, only much higher, and the more daring and venturesome among the population turned their attention to the hunting and scalping of Indians.

Among the most famous of these was Capt. Lovewell, of Dunstable, who repeatedly followed them along the Merrimack as far as this lake, and even penetrated to the country of the Pequaukets. He freely roamed the country of the Winnipesaukees, and with his valiant band traversed the entire circumference, and crossed and recrossed the known territory of this tribe of the famous, once powerful, and dreaded Pennacooks. Seeking indemnity for the past and security for the future, they traversed the wilds of New Hampshire, making this lake on each successive scout, as here was the most probable place to pick up a fresh trail.

That Capt. Lovewell, who explored this lake country, was a bold, energetic, and determined man, there can be no doubt, but whether his skill and prudence as an Indian hunter warrants the famous reputation which he seems to have left behind, is quite another question. It is true he led his men to a great success in the capture of a party of ten Indians in what is now the town of Wakefield, annihilating the entire band; but this achievement was not accomplished at any risk, for it was like shooting a wolf in a trap. However, taking the scalps, stretching them on hoops, and elevating them on poles, they marched proudly into Boston, and received their reward in pounds and shillings.

The great success of this party inflamed the passion for adventure and for gain. Many believed, after the result of this expedition became known, that the Indians had lost their courage, their cunning, and their caution, and supposed that it would be like shooting rabbits, only more profitable; and reflecting that the bounty was, all things considered, equal to the profits on an acre of corn, besides exchanging labor for sport and recreation, they flocked to the standard of the Indian-hunters in considerable numbers. On his second expedition, Captain Lovewell seems to have met with very different success. His bravery, and that of his men, proves that he could stand against overwhelming odds; but it seems that he must have lacked the essential elements of success in a contest with such a wary and cunning adversary, — prudence and caution.

That he forfeited his life, and that of most of his men, does not qualify his rashness, or want of reasonable prudence. All it proves is, that, taking his life in his hand, and trusting to what he considered his qualifications, he was willing to take the responsibility of the position. He risked all, and lost all. However, he taught the Indians a wholesome lesson, for although they were not defeated and routed, still they received a punishment so severe as to alarm them for their security anywhere on this side of the St. Francis country. Other and stronger motives to be sure actuated and influenced the inhabitants in their exterminating warfare against the savages. Few of them but had suffered in the interruption of the labors absolutely necessary for the sustenance of the family, and very many had been bereft of relatives and friends under the most atrocious circumstances, and, feeling as a man whose sheepfold had been invaded by ferocious animals, only that the feeling was intensified by the difference between the loss of friends and property, they determined the beast should be hunted down.

Knowing, by sad experience, something of the baser characteristics of the savage, it may not be improper or untruthful to observe, that the North American Indian possesses, and often displays, in his nobler nature, those higher excellences and sublime qualities which adorn and embellish the human character. Were it possible to reduce him to a homogeneous condition, and illuminate the dark recesses of his soul with the genial and germinating rays of a Christian light, no race could exhibit the elements and emotions of such a condition more gratifying than he. But this can never be accomplished. He can be hunted, pushed, exterminated, but never civilized. It is repulsive to every element of his nature and aspiration of his soul. Unlike the negro he can never be enslaved, and, also, unlike that race, he displays a dignity and gravity in the height of prosperity, or the extremity of adversity, truly refreshing and worthy of imitation. If he is treacherous and cruel to his enemies, he is true and kind to his friends. He is not of that school who think that a good turn *deserves* another; but shows that he appreciates it, and realizes that it merits a return; and his memory is as equally retentive of a favor as an injury. Neither will be forgotten. But he can even be brought to overlook injuries when he realizes that he has no power to resent them, which is a rule with all men. That he really possesses this trait in an eminent degree, Wonnalancet, son and successor of Passaconaway, is a conspicuous example. Having embraced the Christian religion under the teachings of the Apostle Elliot, this noble son of the forest endured abuse and ill-treatment from those whose faith he had adopted, that might have led more considerate people than unlettered savages to doubt the sincerity, value, or importance of such professions: still, with his mind imbued with the true spirit of Christianity, never wavering, he endured all, suffered all; and, though recently redeemed from paganism, his example was, such as the enlightened might profit by imitating.

Even in the darkness of barbarism, in the gloom of unmitigated ignorance, the Indian recognized an overruling Providence; and for success in the chase, on the war-path, or fishing, for bountiful harvests of maize and vegetables, for all the good received, he never failed to offer up to the Great Spirit his grateful acknowledgments: perhaps too demonstrative to suit the tastes of delicate and polished Christians, but it was his way, and who shall say it was not as acceptable as the most studied and eloquent prayer? Having been educated by his surroundings, he was entirely a child of nature, and as such, his appreciation and enjoyment of the sublime and beautiful, which was his gospel, would compare favorably with more fortunate mortals.

His favorite haunts were about the high towering mountains, by the beautiful waterfall, along the bright rolling river, or the great lake whose placid bosom mirrored the celestial hunting-ground of his immortality.

Here, and from scenes like these, he drew his inspiration, and the instinctive generosity and nobility of his impulses proves conclusively that he was susceptible to, and had profited by, the lessons spread out before him on these pages of the great book of Nature. Such was the natural Indian. True, he could never be a Caucasian. Still, except for the demoralizing contact of civilization, for the wrongs and crimes perpetrated upon him, for the bad examples placed before him, and for unjustly seizing and occupying his territory, compelling him to quitclaim the land of his birth and the graves of his fathers, or die in the attempt to maintain his right, — had it not been for practices such as these, the red man would not have been the treacherous, cruel, revengeful fiend of our ideal Indian, but might have stood forth, a character, native and untrained, displaying qualifications which it would have honored more pretending men to copy.

The great, numerous, and powerful Pennacooks, where are they? Two hundred years have effaced every vestige of the race; they are rubbed out like a chalk mark on a black-board; every trace of the blood is obliterated; no scion remains; they have withered as the grass beneath the pavement, and the places that knew them once shall know them no more forever. The few fragile and broken remnants of the race, dispirited, and dimly realizing their ultimate doom, long since turned their backs on old familiar scenes, on the conqueror, and their faces to the setting sun, where year by year his domain is curtailed, and himself more closely environed, until, at no very distant day, he will be totally and finally obliterated from the face of this broad land, and become as much of a myth or tradition, as the centaur, the mastodon, or the sphynx.

To the north and east of Lake Winnipesaukee the view is grand beyond description. Almost from the lake shore the mountains rise high and still higher, not in regular range or column, but scattered promiscuously around, the picket of the main body of great mountains behind them.

Centre Harbor at the head of the lake is the objective point of several lines of mountain travel, radiating from Boston and concentrating here, it being the starting-point for all the stages from the lake to the mountain region.

During the summer the tide of mountain travel is at its flood on this route, and it is estimated that from one thousand to twelve hundred per week are carried both ways in stages.

Attracted by the famed and romantic beauty of the scenery along this route, the tourist and comfort seeker, who possess a taste for the grand and beautiful in nature, will not fail to either go or return by this route, and thus feast their higher sense on the gorgeous panorama that stretches from the foot of the lake to the summit of Mount Washington. Though sometimes encompassed by heat, dust, and discomfort, the compensation is ample, the satisfaction complete.

Centre Harbor is widely known as a summer resort. Travellers reach this place, from New York, via Sound, Norwich, Worcester, Nashua, Concord, and the Weirs. From this point, a short steamboat ride, unequalled for the variety and splendor of natural scenery, and Centre Harbor is reached. From Boston, via Lowell, Lawrence, Concord, and the Weirs, or Portsmouth, Dover, Cochecho Railroad, to Alton Bay; thence, by steamboat, to Wolfboro' and Centre Harbor. The stage road is wonderfully level a large share of the way; still there are some sharp hills which tax the team and tire the passengers, but cool, refreshing zephyrs fan the cheek, and the mind is so engrossed with the wild and remarkable scenery, that the journey is robbed of all toil, and is regarded as a delightful excursion. The route lies through Moultonborough, around the immediate base of Red Hill, — so called, from its appearance at a distance.

On the south side of Ossipee Mountain, which extends into this town, there is a cool and copious spring, impregnated with sulphur and iron. Near the summit of the mountain is a remarkable spring, fifteen feet in diameter, which emits a dozen jets of water to the height of two feet, containing small quantities of fine white sand, and discharges a considerable brook; receiving many tributaries, it becomes, in the course of a mile and a half, a foaming mountain torrent. At that distance from the spring, and not far from Sulphur Spring, it breaks into a broad and furious cascade for fifty feet, then takes a perpendicular plunge of seventy feet.

Ossipee Falls are magnificently grand of themselves, and picturesque in all their surroundings, and it is believed that a summer hotel would be accommodating and remunerating. Near these falls there is a cave, where charred wood and other indications of its once having been a resort for Indians have been found. In 1817, a huge skeleton of a man, supposed to be seven feet high, or more, was found buried in the sand. In 1820, on a small island in the lake, a rusty and ruined gun-barrel, of peculiar workmanship, was found embedded in a pine-tree sixteen inches in diameter. The Ossipee Indians lived about this region, and there is, or was recently, a tree on which was carved the records of the tribe.

"Red Hill, on the north-west side of the lake, is perhaps the very best position from which to obtain a view of the lake and its surroundings. This hill, situated near the lake, rises abruptly two thousand feet, with nothing to obstruct the view.

"Scarcely a stone's throw from the summit is the little Lake Squam, its waters clear as crystal, and sprinkled with green islands, — some of them no wider than a small grass-plot, — some spreading out into fields and pastures, with hills that send forth many a rivulet into the bosom of the lake. Ascending towards the summit of the mountain, the trees, unlike those on the White Mountains, which are gnarled and stunted, appear slender and graceful, and seem to stand for ornament amidst the blueberry and sweet fern, which bear their fruit and fragrance almost to the mountain's top. For weeks, the traveller may daily and hourly discover new attraction in these sweet abodes of nature. To-day, a clear atmosphere presents everything in the brightest hues and charms the mind with the distinctness of every object. To-morrow, a change of atmosphere lends to everything a change of hue, and flings over all a new enchantment.

"Nothing can exceed the splendor of sunrise on this mountain in a calm summer's morning. The stillness of the place; the placid serenity of the waters; the varying positions of objects, as the morning mists rise, and change, and pass away before the sun, now brooding low on the waters, now sailing slowly over the islands, and wreathed in ever-varied forms around their green promontories, — these and other features present to the mind a landscape abounding in that wild beauty which exists where art has not usurped dominion over nature.

"Here, some bright basin is seen to gleam, and anon the eye catches some islet half veiled in mist, and reddening with the first blush of morning. Sometimes, by a pleasing delusion, the clouds become stationary, and the island itself appears to move, and to be slowly receding from the veil of mist. The eye dwells with delight on the villages of the wide country, and the hundreds of farms and orchards which adorn the whole extent of the landscape.

"The fertile islands of the lake are scattered, as if to delight the eye; and when clothed in the deep green of summer, or waving with luxuriant harvests, they seem like floating gardens, immersed in the water. The hills and woods, the shores and eddies, the coves and green recesses, the farms and houses, sometimes retiring from the waters, sometimes approaching to the margin of the lake, — all form a picture formed for the lover of nature to linger and dwell upon with varied and ever new delight. The course of the lake winds at last, and is lost among the distant mountains." *

It is no discredit to any author to observe that the sketch is incomplete. He who has the deepest sense of the grand and the beautiful in nature's wonderful perfection, must realize most sensibly the inadequacy and imperfection of mere words to portray the picture properly.

The true course is to visit, personally, this delightful region. Improved health will compensate the cost, and a lavish and boundless display of nature's wealth will delight the eye, and deeply interest and charm the mind.

Sandwich, the next town, is very mountainous. Sandwich Mountains, a lofty range, beginning in Holderness, extend into Waterville, thence south, Chocorua Peak, in Albany, being their termination, forming two sides of a square.

Bearcamp River has its source in these mountains. Hon. John Wentworth, familiarly known as "long John Wentworth," was a native of Sandwich.

On the summit of a hill, only forty or fifty yards in height, and about the same distance from the stage road, is a very noticeable natural curiosity, which is known as the Lion Rock. This rock attracts the notice of even the most careless observer, while those who journey to the mountains for what they can see, and are sure

* Barstow.

to bestow attention on nearly everything that is visible, never fail to greet this singular curiosity with exclamations of surprise. This rock is so formed and poised that it presents a striking resemblance to a lion, passive. The features, the large head, neck, heavy shoulders, the natural curves and outlines of the body are readily traced. Of the wonderful natural scenery of northern New Hampshire, this rock is a marked feature, and is justly entitled to a place in the famous group located in this region.

Ossipee Mountain, which towers high on the right, has given its name to a powerful gunboat in the U. S. Navy.

CHAPTER V.

Tamworth. — Quakers. — Albany. — Chocorua. — Madison. — Mines. — Conway. — The Notch. — The Willey Family. — White Mountain Railroad.

TAMWORTH is a most decidedly uneven township, surrounded by high mountains. Its people are industrious, enterprising, and thrifty. Apparently hemmed in by impassable barriers on all sides, the traveller, on reaching here, composes himself for a period of quiet, rational, and genuine enjoyment among the hospitable and intelligent people with a certainty of success. Or, if he is so unfortunate as to be destitute of taste for the wonders and beauties of nature, he involuntarily casts about him to ascertain if there is any possibility of escape from this "pent-up Utica."

This town was chartered in 1766, and settled in 1771, by four families, — David Philbrick, Jonathan Choate, Richard Jackman, and William Eastman.

Tamworth Iron Works went into operation previous to 1800. The enterprise was started by a Mr. Blaisdell. Bog-iron ore, taken from Ossipee Lake, was used, and cut nails, anchors, and other heavy articles were made. It is supposed the nails made here were the first that were cut and headed by machinery in the country. These works soon came into the hands of Mr. Nathaniel Weed, a man of great ingenuity, and who, among other things, made a screw-auger, which was said to have been the first implement of the kind ever invented. Not realizing the vast wealth which he could have secured by a patent, he was content with the perfection of his essay. When the Piscataqua Bridge was built, Mr. Weed and many other artisans were employed; taking with him his auger, he proved a very important personage; the old fashioned pod-auger was dispensed with, and relays of hands were appointed, whose business it was to keep the Weed-auger in perpetual motion.

Iron manufacture was here long since abandoned, but mills for various purposes still occupy the site which is on the outlet of Chocorua Lake, — a beautiful sheet of water at the base of the mountain of the same name.

That the pure mountain air, homely fare, and regular habits are conducive to longevity may be said to be demonstrated by the case of a life-long resident of Tamworth, Mr. Stephen G. Philbrick, who was born in Brentwood in 1771, and came to this town to reside the following year. At that time there were but four families; the country was a wilderness filled with catamounts, moose, deer, bears, wolves, etc.

At twenty-three years of age Mr. Philbrick went to Exeter and worked four years at five dollars per month, and one year at six dollars per month. At twenty-seven he married Ruth Rowe, of Kensington, who met with a fatal accident in 1850, being upwards of eighty years of age. He has always lived under the shadow of Old Chocorua, and has never been sick since childhood.

On visiting him, Oct. 11, 1868, he was found husking corn in his barn, which he said he much preferred to idleness. His mind and frame still continue to be wonderfully robust, and all of his faculties are in an excellent state of preservation. He had walked a mile to the neighbors that morning previous to commencing his day's work, which was to husk eight or ten bushels of corn, as that amount, he "reckoned," was "half a man's work."

On being interrogated as to the probable reason of his exemption from sickness, he declared he had never been ambitious to overwork or overplay; that he had been a moderate and industrious rather than a great or spasmodic worker.

Mr. Philbrick was present at the ordination of Mr. Hidden, being at that time above his majority, and was also one of the last surviving four who were present at the erection of the monument on Ordination Rock, and is the last survivor of that venerable quartette. He voted for General Washington for President, and has voted at every presidential election since. His two youngest brothers died in the war of 1812, and his youngest child is approaching the allotted threescore years and ten.

Perhaps the most pleasant portion of the chronicle of this relic of the past is his unswerving and inflexible honesty. Though never a member of any church, his life has been a pattern of morality and uprightness, and it is said that a search-warrant could not produce a single dishonest act in the whole course of his life.

Col. David Gilman of Revolutionary fame was from Tamworth. During the period of his service he was attached to the select military family of Washington. He was a man of gigantic stature, over six and a half feet high, of superior intellectual endowments, dignified bearing, and thorough military air. Having been selected by Washington as one of his most efficient and reliable officers, and despatched on a hazardous and important mission, he met with a serious and painful accident, which obliged him to quit the service. Washington wrote him an autograph letter, accepting his resignation with deep regret, and as a token of his high esteem for him as a "soldier, a man, and a gentleman," the commander-in-chief presented him with his own sword, a fine weapon with solid silver hilt. Col. Gilman returned to Tamworth, and, recovering from his injuries, lived to a very advanced age.

The larger kinds of wild animals, such as bears and deer, still abound in Tamworth, some twenty or more of the former being taken yearly. On the tenth of September last a huge bear was taken west of the residence of Mr. Robert B. Felch, which weighed nearly five hundred pounds, and the rough tallow yielded six gallons of rendered bear's oil. Prior to 1840, wolves were plenty and troublesome, and on one occasion the citizens turned out and killed five in a single day, since which time they have not shown themselves until the present autumn.

This town being mountainous, furnishes many fine trout brooks, the most famous being widely known as Birch Interval Brook, an affluent of the Bearcamp. Its source is three thousand feet above the river; its length about fifteen miles, and it is estimated that one hundred bushels of trout have been taken from it the present season. This brook is large, cold, and transparent, and has many fine waterfalls, on one of which the village is located.

The Sandwich Mountains traverse the north and east border of the town into the adjoining one of Albany, where they terminate in the grim and sullen peak whose angular and rugged summit, destitute alike of soil and vegetation, scarred and seamed by time and the elements, is called Chocorua. This craggy and barren pinnacle, the throne of the thunders, the play-ground of the whispering genii of the winds, whose good spirits sigh mournfully through the firs and spruces which grow so thick in sombre green around its base, or howl and roar about the rocky caverns and naked angles above the line of vegetation, where among the inaccessible cliffs the great bald eagle has his eyrie, rearing its rude crown above the storm-cloud, stands forth conspicuous from the region round, a solitary realm of desolation.

Inaccessible to even the most secret and retiring wild beasts which shun the haunts of man for the unbroken solitude of the wilderness, the eagle rears its young, maintaining unimpaired his "ancient solitary reign."

Athletic men may climb Chocorua; still it requires dexterity and daring, and a scramble to its summit involves danger and fatigue. Reaching a shelving rock or projecting angle, and gaining by this process a little higher footing, or seizing a slender, dwarfed, and doubtful sapling more than liable to be uprooted, taking with it the thin strata of soil upon its root, is no holiday task.

Leave the dogs at home; they cannot follow up, but must be passed from hand to hand, or left behind.

The tribe of the Pequauket had their home about the base of old Chocorua, — their hunting and fishing grounds being the unlimited forest that stretched away to the north, and the Saco and its tributaries which meandered through it; standing, as it did, on the great thoroughfare between the Pennacooks and the Amariscoggins, it was a beacon and a guide visible from afar, even to that other trail away to the west beyond the Pemigewasset which led to the headwaters of the Connecticut and to the St. Francis. Wild and solemn in its annals and aspects, it may not be strange that so fitting a locality should be prolific of legends which still cluster around it.

On an elevated and fertile ridge called "Stevenson Hill," in the western part of the town, and some six hundred feet above the level of the Bearcamp, stands the residence of John M. Stevenson, Esq., which is visible at a great distance. Mr. Stevenson is one of those men who contribute most liberally to the character and prosperity of a town, — public spirited, liberal, and intelligent, carefully surveying and securing its present wants, as well as its prospective interests. His parents were the first couple married by the Rev. Mr.

Hidden, consequently his years, his sympathy, and his interests are intimately connected with its welfare.

He is a living encyclopædia of local and general historic events; and, being possessed of ample means, leisure, and refined tastes, his house is thronged by the educated, and, indeed, all who are favored with his acquaintance, and nothing gives him more pleasure than to furnish conveyances and accompany his friends to all points of interest in this section.

In this secluded, quiet town there is an excellent illustration of the affection and veneration which communities have for good men among them, and the unobtrusive posthumous respect paid to their memory. The good, the wise, and the brave are not forgotten, and in all parts of the State the hand of respect and gratitude for their worthy deeds and lives raises the tablet, unobtrusive perhaps, to perpetuate their memory.

Samuel Hidden was ordained here as a minister of the gospel September 12th, 1792. It being a new town, the people were few and poor. The country was little less than a wilderness, and the people were destitute of a church, or the means of erecting one. Under these unpropitious circumstances Mr. Hidden's ardor was not cooled; his courage was unabated, and his determination to prosecute the good work unshaken.

He was ordained on a large rock seventeen feet high, with an area sufficient to accommodate seventy persons on a level surface, which was reached by a flight of seventeen stone steps. On this modern St. Peter his church was built, — a sure and firm foundation; Ordination Rock typifying the solid, substantial, and weighty measure of his faith, and symbolizing the enduring character of his faithful, efficient, and protracted labors. Desiring to secure a substantial testimonial to his great worth, to obtain a visible memento of his self-sacrificing devotion to the cause he labored so long and so earnestly to advance, a voluntary subscription was raised, to which some of those who had long sat under his faithful ministry were spared to contribute, as well as others to whom his name and works were household words, and a sum was raised sufficient to purchase the land on which this rock stands.

On the 12th of September, 1862, a monument was erected upon this rock, appropriately inscribed, by a grandson bearing his name.

The ceremony of erecting and dedicating the Hidden monument was interesting and impressive. The address, which was eloquent and appropriate, was delivered by E. E. Adams. Four of the residents of the town participated in these ceremonies who had, seventy years before, been present at the ordination of Mr. Hidden. The following extract from a letter written by the Rev. Mr. Coe, of Durham, who was present at the ordination of Mr. Hidden, is very interesting: —

. . . . "I will tell you about the ordination, and yet I know not where to begin or what to say; it defies description. Mr. Hidden was ordained on a large rock, on which fifty men might stand. His foundation must be secure and solid, for the rock will stand till Gabriel shall divide it with the power of God. Early in the morning the people assembled around this rock, men, women, boys, and girls, together with dogs and other domestic animals. It is an entire forest about this place. The scenery is wild. On the north is a high hill, and north of this is the mountain called Chocorua, which touches heaven. On the south and in all directions are mountains steep and rugged. I expected to have heard the howling of the wolf and the screeching of the owl; but instead of these were heard the melting notes of the robin, and the chirping of the sparrow and other birds, that made the forest seem like paradise. The men looked happy, rugged, and fearless; their trowsers came down about half way between the knee and ankle; their coats were mostly short, and of nameless shapes. Many wore slouched hats, and hundreds were shoeless. The women looked ruddy, and as though they loved their husbands. Their clothing was all of domestic manufacture. Every woman had on a clean checkered linen apron, and carried a clean linen handkerchief. Their bonnets! Well, I cannot describe them: I leave them to your imagination. But think of the grandeur of the scene! A great rock the pulpit, the whole town the floor of the house, the canopy of heaven the roof, and the tall, sturdy trees the walls! Who could help being devoted?

"This is the place nature has formed for pure worship. Long shall this rock stand like the rock on which our fathers landed. Long may this church make the wilderness and the solitary place glad, and the desert bloom as the rose.

> "'Father, thy hand
> Hath reared these venerable columns; thou
> Didst weave this verdant roof. Thou didst look down
> Upon the naked earth, and forthwith rose
> All these fair ranks of trees. They in thy sun
> Budded and shook their green leaves in thy breeze
> And shot towards heaven. They stand tall and dark,
> Fit shrine for humble worshipper to hold
> Communion with his Maker.'"

Scattered over Tamworth and the adjoining town of Sandwich are considerable numbers of a peculiar and remarkable sect of Christians, called Friends or Quakers, who have two meeting-houses, and maintain regular public religious worship. Among the earliest Friends who settled in Tamworth was Mr. Parker Felch and family, who came here in 1800, and whose descendants are still among the most faithful and zealous of the Quaker sect, earnestly maintaining the forms and the ancient faith, and proving by their daily walk and conversation that its requirements and its natural tendency is to that approved kind of genuine Christianity which is creditable to the professor and advantageous to the community, and refreshing to contemplate in these degenerate days.

This peculiar and interesting denomination of Christians may be entitled to a brief notice, as their practice seems an exemplification of the moral law and the faith that is in them, as well as a simplification, if not improvement, of the freedom, protection and restraint sought to be accorded and secured by legal human government.

Quakers are sometimes liable to be confounded with Shakers, whom they are totally unlike, except in the plainness and simplicity of demeanor, language, and dress, and in the peaceable precepts of their creed. Like most other sects they countenance, encourage, and believe in the sacred character of the marriage and family relations, in peaceable, honest, and efficient civil government, trade, business, accumulating and devising property, and all which a good citizen may do without detriment to the State or injury to his fellow-man. Persistent and unwavering in adhesion to their faith and practice, they have clung to it with a tenacity which shows the depth of their conviction. They have unflinchingly withstood the wild and furious tempests of religious fanaticism and intolerance which have from time to time swept over them, threatening to obliterate the Friend and his faith together.

Stoutly maintaining the principle of religious liberty, and steadily and boldly, without fear or favor, displaying those outward and visible emblems by which, if in no other manner, they may be known and recognized of men; never propagandists, and singularly unattractive and unfashionable in personal adornments, in their mode and places of religious worship, and in the austerity of their moral requirements, their theory has never been popular or their increase rapid; their proselytes springing from the seed always sown by persecution and violent and unreasoning denunciation rather than from spontaneous attraction of the faith or the forms. Their theory seems to be a combination of civil and religious government, a harmonious blending of Church and State, practical, economical, and efficient. They do not believe in a hireling ministry; logically concluding, that if inspired with power they will also be with the duty, as was the Saviour, to preach from the highways and byways and hill-tops without money and without price.

Litigation among the members meets with unqualified disapprobation, it being engaged in on pain of excommunication. Thus it will be seen that lawyers and ministers among the Quakers raise cattle and potatoes, or engage in some other productive employment. Peremptory regulations and provisions against the employment of paid preachers they regard as not only a moral improvement, but a brilliant stroke of policy, obviating the necessity of harboring and maintaining a big devil in their community, or employing persons at large salaries to battle him.

"Our ancient testimony against an hireling ministry, or any contributions to the maintenance and support thereof, being founded on examples and precepts of our Lord and his primitive followers and disciples; it is the sense of this meeting that friends be careful to support the same, by a faithful testimony against contributing towards their salaries by tax or otherwise, and against the building or repairing their meeting-houses.

"Also this meeting doth advise that if any person professing the truth among us and esteemed a Friend, shall refuse speedily to adjust the difference, or refer it as before advised to (arbitration), complaint be made of that person or persons unto the monthly meeting to which they doth belong, and if after admonition he shall refuse so to refer his case, that then the meeting do testify against such person, and disown him to be of our society until he shall comply with the equal methods and agreements of our society, and by such, his compliance doth declare that he is for peace, and doth seek and desire it.

"We have, as a people, looked upon ourselves as well as the primitive Christians to be included in the notable prophecy of Isaiah ii. 4: 'They shall beat their swords into ploughshares, and their spears into pruning-hooks, and learn to war no more.'" *

It will be seen that war or bearing arms for any purpose is repudiated and forbidden, and, as consistency is claimed to be one of their brightest jewels, they object to the payment of taxes for any such purpose, but suffer them to be collected, owing to their non-resistant principles. Such as are not self-sustaining are provided for by the society, all funds being raised by voluntary subscription; consequently no poorhouses are maintained by them or among them, and squalid poverty is even less frequent among them than immense wealth, and both are rare. Government, temporal and spiritual, — so far as it does not conflict with the rightful authority and jurisdiction of the common law, — is defined and administered by representatives, male and female, who assemble in conventions known as monthly, quarterly, and yearly meetings, and public or Quaker opinion is the sole official relied upon for the enforcement of the decrees or recommendations of the law and gospel givers of the Quaker sect. In their dealings they are said to be scrupulously exact; the impression seems to be that Quaker measure is more than just, it is generous. This is believed to be a serious mistake; it is more than they claim; if they are square even to the last pennyweight, no more, no less, it is not only creditable to their sense of justice and fair dealing, but places them conspicuous, in this regard, for imitation of seared and callous consciences, and the advantages of many purchasers.

There are now about twenty meeting-houses in the State, and though they do not increase in numbers and influence so rapidly as some denominations, still they maintain the even tenor of their way, and may in some respects serve as a profitable example to many other of larger numbers, of more influence, and greater pretensions.

* Discipline.

"While the magistrates of Portsmouth were busy with the witches, religious intolerance broke out fiercely against the Quakers. During the whole period of this persecution, New Hampshire was but an appendage to Massachusetts, and the laws by which Quakers were whipped and led through the streets of Dover tied to carts were laws of Massachusetts. The stain of that vindictive persecution attaches itself to New Hampshire, because she had a small representation in the assembly of Massachusetts when those laws were enacted. The civil authorities of Boston justified their proceedings with the specious pretence of securing the peace and order of society. They declared the 'vagabond Quakers' to be 'capital blasphemers,' seducers from the glorious Trinity, open enemies to government, subverters both of church and State. Accordingly, a law was published prohibiting the Quakers from coming to the colony on pain of the House of Correction; notwithstanding which, by a back-door they found entrance.

"The penalty was then increased to cutting off the ears of those who offended the second time. This barbarous punishment was inflicted in several instances, for which the *public safety* was the ready apology. But even this proved ineffectual; and the offenders were next banished upon pain of death for returning. But this availed nothing; the Quakers returned, and sealed with their blood the testimony of their faith. Of all the wrongs which man has inflicted upon his fellow-man, is there one which has not been perpetrated in the name of religion and for public good?

"On the 27th of October, 1659, Robinson and Stevenson were led to execution, attended by two hundred armed men, besides many horsemen. When they had come near the gallows, a coarse and vulgar priest cried out tauntingly to Robinson, 'Shall such Jacks as you come in before authority with hats on?' To which the martyrs made a mild reply. The prisoners then tenderly embraced each other, and ascended the ladder. When Robinson signified to the spectators that he 'suffered not as an evil-doer,' the voice of the priest was again heard, 'Hold thy tongue; be silent; thou art going to die with a lie in thy mouth.' The sufferers were soon launched off; their last words were silenced by the beating of drums. When William Ledra was brought to the gallows, he made a speech which 'took so much with the people that it wrought a tenderness.'

"Allen, an officious priest, was near, whose business it was to make the martyr odious, and instantly interrupted him. 'People,' cried Allen, 'I would not have you think it strange to see a man so willing to die.'

"The hangman was commanded to make haste with Ledra, and so he was turned off, 'and finished his days.' But his friends, with solicitude, gathered around the foot of the gallows, caught the body in their arms as it fell, bathed it with tears, and, having waited until the hangman had stripped it of the clothes, laid it decently in a coffin. Thus intolerance had another victim.

"When the news of this bloody work was carried to England, and reached the king, an order was forthwith issued to Gov. Endicott to suspend all executions, and send the Quakers to England for trial, — a privilege which they claimed when brought before the courts of Massachusetts.

"The next year, three women were publicly whipped in New Hampshire in the depth of winter; the constables were ordered to strip them and tie them to a cart; then to drive the cart, and whip these three tender women through eleven towns, with ten stripes apiece in each town. The route lay through Dover, Hampton, Salisbury, Newbury, Rowley, Ipswich, Wenham, Lynn, Boston, Roxbury, and Dedham, a distance of near eighty miles. They were whipped at Dover and Hampton, and then carried, 'through dirt and snow half the leg deep,' in a very cold day, to Salisbury, and there whipped again. They would probably have perished long before reaching the end of the route, but at Salisbury they were happily released. Walter Barefoot persuaded the constable to make him his deputy, and having received the warrant set them at liberty, and they returned to Dover." *

"The prosperity of Albany has been retarded by a remarkable disease which almost entirely prevents the raising of neat stock. Its peculiarities are a loss of appetite, costiveness, contraction of the abdomen, followed in a few days by powerful evacuations, by which the animals are rapidly reduced, and soon die.

"Superstition and tradition point to the curse of Chocorua as the cause; but the better supposition is, doubtless, that it is owing to certain properties contained in the water or in the soil. Science will,

* Barstow.

we trust, ere long point out the cause of the evil which so much injures and afflicts man and beast." *

Among the traditions still observed, and that which gave Chocorua its name, is one possessed of considerable interest. Chocorua, a great chief, wily and cruel, having been for a long time a terror and scourge to the whites, it was determined at almost any sacrifice to destroy him. Accordingly a strong force of bold and daring men assembled to hunt him down and exterminate him. Pursuit was at once commenced; himself and his band discovered; attack and pursuit alternated; the struggle was desperate and bloody, but not doubtful. At length, worn down by fatigue, and fearfully decimated by the unerring musket, and alarmed by the unshaken persistence of the adversary, the remnant of his band scattered and fled in every direction.

Surprised and disappointed at this unexpected manœuvre, and fearing the cunning and desperate chief would elude them and escape, they singled him out, allowing the others to go at will, and bent all their energies to his capture or destruction. Like bloodhounds they followed his trail. Doubling, denning, and circling like a fox in his dire extremity, adopting every stratagem to elude his pursuers known to savage fertility, he was as surely unearthed and headed off in every attempt to escape, until, at last, closely crowded, he took to this mountain. Spreading out to avoid any possibility of his escape in any direction, they commenced the toilsome ascent, and gradually and steadily closed in upon the victim. High-spirited, and hating the whites with unmitigated animosity, as well as fearing and despising them, he scorned to ask quarter or any favor whatever, but continued climbing up the rugged steep, nerved by the spirit of desperation. Up, and still up he toiled, higher and to a still more dizzy height; but his pursuers hung like hungry wolves about him; he could gain nothing of them. At length he reached the highest summit, the tip-top pinnacle, and finding he was securely cornered and irretrievably lost, that his enemies were immovably posted around him like a wall of death, he prepared for his doom. Directing his gaze towards the declining sun, with face upturned, he raised his voice to the Great Spirit, and called down his curse upon the land.

* New Hampshire As It Is.

What a picture! Standing on the highest point of this bleak and barren mountain-top, a pagan savage in war-paint and aboriginal trappings, lifting up his heart and voice in supplication to Him who created and governs the universe, while surrounded by a cordon of Christian men, eager for his life and thirsting for his blood (they had previously killed his son), he prayed that the Great Spirit would come down in his wrath and vengeance, and curse the people and the soil, and above all that he would lay his curse upon the horned cattle, so that they should not live and thrive in all this section. Having concluded his prayer, Chocorua turned his eagle eye upon the bright and beautiful world around him, as if to enjoy one long, lingering, parting look at the scenes familiar and beloved; he then turned his face to the foe, and with a defiant gesture and a horrible yell of combined and concentrated rage and triumph, waking the echoes, as if it afforded him a grim and ghastly satisfaction to rob his relentless and vindictive pursuers of the prize they coveted, he bounded high in the air, and cleared the brink of the frightful precipice, — down, down over the jagged rifts and projecting angles of this rough and perpendicular bluff, for two thousand feet, landing a shapeless mass of flesh, on which the eagle gorged her young. Brave old Chocorua, determined and desperate, his courage availed him nothing; and though the story of his death still survives, even the fragments of his bones, bleached and decayed by time and the elements, were long since reduced to dust and mingled with the very soil he cursed. Superstitious people believe that this prayer was heard and answered; for it is a fact that since then a very destructive and fatal disease has constantly prevailed, exactly the same disease from year to year, unknown elsewhere, destroying many dollars' worth of cattle annually. Animals that are driven from this to other sections never have it, and the disease is known in all the region round about as the "Albany ail." Tourists and strangers, who have seen the disease, aver that it appears much less wholesome than an article which goes by a similar name, that they have met with elsewhere.

There are many Indian legends current among the inhabitants of the territory about the base of old Chocorua; some, no doubt, founded on fact; others, probably, inventions of the marvellously inclined. Gray-haired men still recount the hairbreadth escape from, or the miraculous frustration of the Pequaukets, or the merciless cruelties of Chocorua, Paugus, and their compeers, to the more recent comers, or knee-high youngsters around the cheerful log fire, who with glistening eyes, and their hearts in their throats, imagine that every sound betokens the approach of these monsters, every sigh of the night-wind is a war-whoop that the neighbor who steps in to "look at the paper a moment," or to "borrow a tallow-dip," is an Indian in disguise; that one is secreted behind each tree; that the outside door is a sally-port through which he momentarily expects to see the plumed and painted braves rush in pell-mell, and involuntarily feels for his scalp.

About twenty miles from Conway, by the course of the river, on the Swift River interval, there is a settlement of some dozen productive and valuable farms, producing lumber, cattle, hay, cereals, potatoes, etc., but no corn, the seasons being too short. One of the farmers said they never had more than two months in the year without frost, rarely ever more than one. These people, if they require such luxuries, must travel twenty miles over an indictable road, for the doctor, the post-office, the grocery, or the church, notwithstanding which they are as moral, intelligent, healthy, and well-fed, as more fortunate people. From this settlement, by the road, it is fifty miles to the Willey House in the White Mountain Notch, three miles south of or below the Crawford House, while an indescribably beautiful and pleasant march of less than ten miles, by a blazed or spotted line of trees, through a portion of Hart's and Sawyer's locations, toilsome and tiresome of course to those unaccustomed to forest and mountain tramps, brings one to the same point; while, turning in the other direction, a tramp of some seven miles across the Sandwich range, by no means a difficult undertaking, and Tamworth is reached, the distance by the road being a round fifty miles.

From the summit of old Chocorua, a splendid view is obtained. A number of villages, like clusters of bird-cages ranged upon a green lawn, are distinctly seen. Mountains, huddled together in inextricable confusion, resembling gigantic tumbles and winrows of hay in an extensive meadow, many lakes and ponds, and several rivers are seen through stretches of intervals and forest, like silver threads in an emerald cloth, while the great forests of the deciduous and the perennial in every direction stretch away interminably into the blue and misty distance. If the atmosphere is clear, it well repays the time and toil; otherwise it is unmitigated wretchedness. Clouds, thick and moist, roll far below the summit; the wind howls, and a chilling torrent of rain descends. Nothing can be seen, and nothing is to be gained except the friendly shelter of the place from whence the excursionists set forth, which, if accomplished without soiled and rent garments, or bruises, is a decided success, a masterly retreat.

Along this range several peaks are named. Among these, Toadback, a singular-shaped and densely wooded mountain; Whiteface, high and bald, and Mount Israel. Whiteface derives its cognomen from the fact, that its south-western face, from the summit far down the slope, is a vast bed, comprising hundreds of acres of crystallized granite quartz. These crystals are uniformly hexagonal, and are a study for the geologist.

From the tops of these mountains, especially Chocorua, a grand view is obtained to the westward beyond the Pemigewasset country, southward across the beautiful Winnipesaukee, while to the east is the wild valley of the Upper Saco, and rolling rapidly at its very base is that fork of the above-named river, which is so well described by its name, the Swift Branch; uniting with the Pequauket at Conway, it forms the Saco. The high and wild locality of their source up in the heart of the White Hills, and the romantic career of these forks of the Saco, can be traced for a long distance, as well as the beautiful river itself, below their confluence. This mountain has also given its name to a gunboat in the navy.

Madison was formerly included in the territory of Eaton, but the town was divided, and this portion of it received the above name. Madison Brook is familiar to experienced trout-catchers, and is a great resort for this class of sportsmen. In this town there is a mine which was first discovered by the outcropping of the ore. A company was organized, buildings erected, and the mine opened. The company was incorporated as the Carroll County Silver Mining Company; but the attempts to separate these ores proved ineffectual, and the company failed. Recently, however, a new company has control of it, at the head of which is Henry J. Banks, Esq., a man of force and enterprise. Recent experiments seem to have demonstrated the feasibility of separating the ores by specific gravity.

These minerals are found in a combination of granite, which, being separated from the metals, yield seventy per cent. of crude ore, the lead and zinc being also separated by the same process. The lead yields one thousand four hundred pounds to the ton from the crucible of pure metal, and one hundred ounces of pure silver, and the zinc gives an average of fifty-nine per cent.

The mine is believed to be inexhaustible, and rather increases in purity than otherwise as the operations of the miners are extended. This is, without doubt, the richest mine of argentiferous ore yet opened in the State, it being not only remarkably rich, but apparently inexhaustible, and should the experiments now being made result in the discovery of a cheap, rapid, and efficient method of separating the several valuable minerals from each other, this mine will prove not only a source of wealth to its fortunate proprietors (Messrs. Banks, French, and Butler), but an important item in the products of New Hampshire. Mr. Banks is proprietor of the well-known hotel at West Ossipee. This hotel is surpassed by few even in our large cities for the convenience and comfort of its internal arrangements. The rooms are supplied with pure cold water from a mountain spring, and its location by the Bearcamp renders it cool, pleasant, and comfortable for the permanent or transient sojourners; while the scenery in the immediate vicinity of the house is charming. The distant view is peculiarly grand.

Conway seems to be the grand gateway to the White Mountain region. Located on the territory of the Pequaukets, it is romantic and delightful as a summer resort, and wealth and taste have been united to enhance its natural advantages in this regard. It is a lovely and important inland town. The Pequauket and Swift Rivers unite here and form the Saco.

In 1765 Daniel Foster obtained a grant of the territory. The conditions were that each grantee should pay an annual rent of one ear of corn for ten years, if demanded.

On the south side of Pine Hill is a huge block of granite, believed to be the largest fragment of rock in the State. Magnesia and fuller's earth are found here. The Saco, having sources among the mountains, is subject to a very rapid rise, and has been known to rise thirty feet in a single day. At the west end of Ossipee Lake there is a mound of earth some fifty feet in diameter, and ten or twelve feet high. It is artificial, and from it have been taken several entire human skeletons, also tomahawks and other relics, supposed to be of the Pequankets, and it is believed this mound was one of the principal burial-places of the tribe.

From Conway to the Crawford House the thoroughfare is, perhaps, as well known and as much travelled as any section of road of its length in the country. Throughout the journey new objects of attraction constantly burst upon the view. Kearsarge, in duplicate, which should have been called Pequauket, is unquestionably the most symmetrical mountain in the State. Rising two thousand seven hundred feet above the plain at its base, its sides taper gradually to the summit. The uncommon regularity of its outline could be properly attributed to artificial causes.

The most famous gunboat in the American navy bears the name of Kearsarge. It was this powerful ship which boldly grappled with that famous scourge of the seas, the Alabama, and, by a well-directed or lucky shot, ruined her future prospects.

Moat Mountain never fails to attract the notice of all. Goodrich Falls, picturesque and grand, are a great point of attraction, while the silver cascades elicit from every one exclamations of unbounded admiration and delight.

A large rock by the wayside bears the name of an heroic but unfortunate girl who perished in the snow near it in an insane attempt to follow her faithless lover; and, to this day, the superstitious who pass this spot in the lonely hours of night imagine his perturbed spirit hovers around and moans sadly, as if in perpetual torment.

The Willey House, where the great slide occurred, arrests the attention and progress of all, and the scene of the catastrophe is surveyed with mingled emotions of interest and pity for the sad fate of the unfortunate Willey family.

"A few weeks after the day that I passed in this secluded valley there was a violent storm of rain, which produced an avalanche, or, according to the language of that region, a *slide*, and the family of Mr. Willey, — by whom I and my companions had been entertained during our journey, — consisting of himself and wife, five children, and two hired men, were buried beneath the rocks, trees, and earth that were borne down by the freshet.

"It was supposed that they were alarmed by the noise, and left the house in their flight, and thus met inevitable death. Just before it reached the house, the avalanche divided into two parts, one passing each side of the house, leaving it untouched. Thus was stricken from the face of the earth a group, which the virtuous and the happy could not but admire, which the rich and the proud might envy. No mortal eye was permitted to witness and survive the agonies of that awful moment; no mortal ear caught the expiring groan of the sufferers. The horrors of the catastrophe are imprinted on the memory, of no child of earth; yet were the hairs of their heads all numbered; and who is there that would not admire the kindness of that Providence which left no bruised reed standing amidst a scene of bereavements; no parent to weep over the mangled and faded flower; no infant bud cut from the parent stock to wither and die in the blast?" *

The scene of this appalling calamity continues to be invested with a deep and mournful interest. It has been visited, in person, by very many, — more than one hundred thousand people. Every account of it is read with avidity, and no description of it is perhaps more interesting than the simple, touching, and beautiful ballad, "The Willey House. A Ballad of the White Hills; by Dr. T. W. Parsons, of Boston": —.

"Come, children, put your baskets down,
And let the blushing berries be;
Sit here and wreathe a laurel crown,
And if I win it, give it me.

"'Tis afternoon, — it is July, —
The mountain shadows grow and grow;
Your time of rest and mine is nigh, —
The moon was rising long ago.

"While yet on old Chocorua's top
The lingering sunlight says farewell,
Your purple-fingered labor stop,
And hear a tale I have to tell.

"You see that cottage in the glen, —
You desolate, forsaken shed, —
Whose mouldering threshold, now and then,
Only a few stray travellers tread.

---

* Extract from the "Boston Galaxy," 1826, Hon. Joseph T. Buckingham, Editor.

"No smoke is curling from its roof,
  At eve no cattle gather round,
No neighbor now, with dint of hoof,
  Prints his glad visit on the ground.

" A happy home it was of yore;
  At morn the flocks went nibbling by,
And Farmer Willey, at his door,
  Oft made their reckoning with his eye.

" Where yon rank alder-trees have sprung,
  And birches cluster thick and tall,
Once the stout apple overhung
  With his red gifts the orchard wall.

" Right fond and pleasant in their ways
  The gentle Willey people were;
I knew them in those peaceful days,
  And Mary — every one knew her.

" Two summers now had seared the hills,
  Two years of little rain or dew;
High up the courses of the rills
  The wild rose and the raspberry grew.

" The mountain-sides were cracked and dry,
  And frequent fissures on the plain,
Like mouths, gaped open to the sky,
  As though the parched earth prayed for rain.

" One sultry August afternoon,
  Old Willey, looking towards the west,
Said, ' We shall hear the thunder soon;
  Oh! if it bring us rain, 'tis blest.'

" And even with his word, a smell
  Of sprinkled fields passed through the air,
And from a single cloud there fell
  A few large drops, — the rain was there.

"Ere set of sun a thunder stroke
  Gave signal to the floods to rise;
Then the great seal of heaven was broke;
  Then burst the gates that barred the skies;

" While from the west the clouds rolled on,
  And from the nor'west gathered fast, —
' We'll have enough of rain anon,'
  Said Willey, ' if this deluge last.'

"For all these cliffs that stand sublime
 Around, like solemn priests appeared,
Gray druids of the olden time,
 Each with his white and streaming beard.

"Till in one sheet of seething foam
 The mingling torrents joined their might;
But in the Willeys' quiet home
 Was naught but silence and 'good-night.'

"For soon they went to their repose,
 And in their beds, all safe and warm,
Saw not how fast the waters rose,
 Heard not the growing of the storm.

"But just before the stroke of ten
 Old Willey looked into the night,
And called upon his two hired men,
 And woke his wife, who struck a light;

"Though her hand trembled, as she heard
 The horses, whinnying in the stall;
And 'Children' was the only word
 That woman from her lips let fall.

"'Mother' the frighted infants cried,
 'What is it? has a whirlwind come?'
Wildly the weeping mother eyed
 Each little darling, but was dumb.

"A sound! as though a mighty gale
 Some forest from its hold had riven,
Mixed with a rattling noise like hail!
 God! art thou raining rocks from heaven?

"A flash! O Christ! the lightning showed
 The mountain moving from his seat!
Out, out into the slippery road,
 Into the wet with naked feet! —

"No time for dress, — for life! for life!
 No time for any word but this;
The father grasped his boys; his wife
 Snatched her young babe — but not to kiss.

"And Mary with the younger girl,
 Barefoot and shivering in their smocks,
Sped forth amid that angry whirl
 Of rushing waves and whelming rocks.

"Far down the mountain's crumbling side,
　　Full half the mountain from on high
　Came sinking, like the snows that slide
　　From the great Alps about July.

" And with it went the lordly ash,
　　And with it went the kingly pine,
　Cedar and oak, amid the crash,
　　Dropped down like clippings of the vine.

" Two rivers rushed, — the one that broke
　　His wonted bounds and drowned the land,
　And one that streamed with dust and smoke, —
　　A flood of earth, and stones, and sand.

" Then for a time the vale was dry,
　　The soil had swallowed up the wave;
　Till one star looking from the sky,
　　A signal to the tempest gave.

" The clouds withdrew, the storm was o'er,
　　Bright Aldebaran burned again;
　The buried river rose once more
　　And foamed along his gravelly glen.

" At morn the men of Conway felt
　　Some dreadful thing had chanced that night,
　And some by Breton woods who dwelt
　　Observed the mountain's altered height.

" Old Crawford and the Fabyan lad
　　Came down from Ammonoosuc then,
　And passed the Notch — oh! strange and sad
　　It was to see the ravaged glen.

" But having toiled for miles, in doubt,
　　With many a risk of limb and neck,
　They saw, and hailed with joyful shout,
　　The Willey House amid the wreck.

" That avalanche of stones and sand,
　　Remembering mercy in its wrath,
　Had parted, and on either hand
　　Pursued the ruin of its path.

" And there, upon its pleasant slope,
　　The cottage, like a sunny isle
　That wakes the shipwrecked seaman's hope,
　　Amid the horror seemed to smile.

"And still upon the lawn before,
    The peaceful sheep were nibbling nigh;
But Farmer Willey at his door
    Stood not to count them with his eye.

"And in the dwelling — O despair! —
    The silent room, the vacant bed!
The children's little shoes were there, —
    But whither were the children fled?

"That day a woman's head, all gashed,
    Its long hair streaming in the flow,
Went o'er the dam, and then was dashed
    Among the whirlpools down below.

"And farther down, by Saco's side,
    They found the mangled forms of four,
Held in an eddy of the tide;
    But Mary, she was seen no more.

"Yet never to this mournful vale
    Shall any maid, in summer time,
Come, without thinking of the tale
    I now have told you in my rhyme.

"And when the Willey House is gone,
    And its last rafter is decayed,
Its history may yet live on
    In this your ballad that I made."

The gateway of the Notch is twenty-two feet wide, and is for miles a narrow gorge, scarcely wide enough for the road and the Saco River, which foams through the Notch, a rapid torrent.

From the mountain-side comes tumbling a magnificent cascade, more than eight hundred feet high, directly to the road and river. Another, not so high, falls over three several precipices, over the last of which it divides into three separate streams, uniting again at the foot of the mountain.

Ascending the mountain, a plain is reached far up, which is the base of the final pinnacle, towering fifteen hundred feet; still higher, a ragged, barren summit, where, in clear weather, the most extensive prospect opens, to be found east of the Rocky Mountains. The vision extends from the deep blue of the broad Atlantic to the deep green of the Vermont hills; from the Canadas on the north to the old Bay State on the south, presenting a view, in extent and variety, unequalled.

Long years before the white man set foot upon the soil of New Hampshire, the red man had an unwritten history of all this mountain region, founded on his superstitious veneration for all that is sublime, beautiful, or grand in nature. The whole mountain region of northern New Hampshire he called by the general name of Agiochocook. They had a tradition that a deluge destroyed all the people, except one Powow, — the traditional Noah of the red man, — and his squaw, who, fleeing before the rising waters, finally saved themselves upon these heights, which were inaccessible to the floods, and finally repeopled the earth.

"These awful summits they regarded with superstitious veneration. The red man believed that a powerful genius presided on their overhanging cliffs, and by their waterfalls. His imagination peopled them with invisible beings. He saw the Great Spirit in the clouds gathered around their tops; he heard his voice, speaking in the revels of the storm, and calling aloud in the thunders that leaped from cliff to cliff, and rumbled in the hollows of the mountains.

"A god resided in the stars, the lakes, and the recesses of the grottos. He saw him in the clouds, and heard him in the winds — frowning in the wintry blast — breathing in the zephyrs of spring — smiling in the first blush of morning, and the last hue of twilight that lingers above the pines in the western sky.

"Influenced by fear, the Indians never ascended the White Mountains. They supposed the invisible inhabitants would resent any intrusion into their sacred precincts.

"But the emotions of the white man were very different. He, especially if he is a Yankee, 'wants to know;' and more than two hundred years ago he explored these solitudes, and his report, though different from the tradition of the Indians, was quite as exaggerated. But the Indian is known no more among these wild, romantic scenes; the fearless and enthusiastic explorer of the early time has, also, long since gone, and his reports, largely drawn from imagination, conjecture and fancy, have long since, like the tradition of the red man, been dispelled by a better knowledge of the region, by the light of facts and science; and though the fleecy vapors still whirl about these awful peaks, and the winds moan like uneasy spirits from

Pandemonium, and Jove still sits enthroned upon the lofty pinnacles, and milky torrents still roll down their sides, there is nothing of mystery hovering about this section now, excepting the inexplicable mystery of their creation.

"In 1642, Capt. Neal explored the White Mountain region, moved to this enterprise, doubtless, by a passion for discovery and adventure, and there could be no other place so well calculated to gratify a disposition of that kind, as the unknown and mysterious region which included these great mountains.

"Such an impression had they made upon the imagination of Neal, that he set out on foot, attended by two companions, to reach them through an unexplored forest. He described them, in the most exaggerated style, 'to be a ridge extending an hundred leagues, on which snow lieth all the year, and inaccessible except by the gullies which the dissolved snow hath made.' On one of these mountains the travellers reported 'to have found a plain of a day's journey over, whereon nothing grows but moss; and, at the further end of this plain, a rude heap of mossy stones, piled upon one another a mile high, on which one might ascend from stone to stone, like a pair of winding stairs, to the top, where was another level of about an acre, with a pond of clear water.' This summit was said to be far above the clouds, and from hence they beheld a vapor like a vast pillar drawn up by the sunbeams out of a great lake into the air, whence it was formed into a cloud.

"The country beyond these mountains, northward, was said to be 'daunting terrible,' full of rocky hills, as thick as molehills in a meadow, and clothed with infinite thick woods. They had great expectations of finding precious stones; and something resembling crystals being picked up was sufficient to give them the name of 'Crystal Hills.' From hence they continued their route in search of a lake, and 'faire islands.' But their provisions were now well-nigh spent, and the forests of Laconia yielded no supply. So they were obliged to set their faces homeward, when 'the discovery wanted one day's journey of being finished.'" *

Late in the year, depressed with that disappointment which ever treads upon the heels of extravagant expectations, they returned from their melancholy journey across the wilderness.

* Barstow.

Extract from a letter of George B. Roberts, Esq., September, 1868: —

"Here is the gateway between the Saco and Ammonoosuc valleys, and we enter the latter, riding four miles to the Fabyan or Mount Washington stand, where work has already commenced for the erection of one of the finest houses in the State. This is the starting-point of a turnpike road, six miles in length, to the depot of the Mount Washington Railroad, and taking us into a regular amphitheatre of hills, asking of us, it is true, a heavy toll, but amply repaying, by keeping us in full view of the highest summits, and grandest elevations of land this side of the Rocky Mountains, and seems destined to be the great thoroughfare to 'Tip-top.' The only other approach which equals this in beauty and sublimity is from the Crawford House, the bridle-path over Mounts Clinton and Pleasant; but I wish to say something of the Mount Washington Railroad, which traverses the western side of the mountain of the same name.

"This road, which is purely the invention of Mr. Marsh, of Littleton, and to which he has devoted several years of toil and much of his private fortune, is not yet complete, is carrying passengers from base to summit, and back, for two dollars each, the distance being about two and one-half miles, and the elevation more than four thousand feet. When the road is completed, the average grade will be about five feet per rod. It has upon its track two engines of forty horse-power each, and when not engaged in taking passengers, are carrying lumber to the top of the mountain to finish the road. The company have a saw-mill at the depot, which was first started as a water-mill, using the 'Tyler wheel,' but the water was found to be insufficient at some seasons of the year, and a steam engine of thirty horse-power was substituted, which runs a fifty-inch circular saw, and is cutting lumber at the rate of ten thousand feet per day. Much of the road is built on trestle work, requiring a large amount of lumber. There is nothing about the running wheels of the engine or cars, or the rails upon which they run, which is different from the other roads, but the road has a centre rail, and each engine and car two driving and holding wheels, the rail upon the plan of the segment, the wheels on the plan of the gear, completely fitting the segment rail, and all fitted with powerful brakes,

— any one of which will instantly stop an engine and two loaded cars upon the steepest grade. This is easily done, as the cars move only about two miles per hour, and the breaking of a wheel or axle would result in no serious accident. The road has been partially in operation for the last year, and fully meets the most sanguine expectations of its inventors and constructors. The train upon which our party ascended, carried thirty gentlemen and ten ladies, and all seemed highly delighted with the ride.

"This road, and its practical operation, of course settles the question of the practicability of ascending mountains by steam, and too much praise cannot be awarded Mr. Marsh, for overcoming obstacles and surmounting difficulties which would have wrecked the fortune of a less enterprising and persevering genius."

CHAPTER VI.

Gilford. — Alton. — Wolfboro'. — Tuftonborough. — Meredith. — Sanbornton. — Northfield. — Canterbury. — Shakers. — Pembroke. — Suncook River. — Gilmanton. — Barnstead. — Pittsfield. — Epsom. — Allenstown. — Contoocook River. — Hillsboro'. — Gov. Pierce. — Henniker. — Washington, etc. — Hopkinton.

GILFORD, on the lake shore, is a flourishing town, enterprising and lively. The principal village is on the falls of the Winnipesaukee River, and is largely engaged in manufacturing. The extensive repair-shop of the Boston, Concord, and Montreal Railroad is located here, giving profitable employment to a large amount of superior mechanical skill. Besides this and mills, there are many factories and shops for various manufacturing and mechanical purposes. Several large islands in the lake belong to Gilford. Gunstock Mountains, a range of considerable elevation, rising in Gilmanton, extend into this town almost to the lake. Gilford was originally a part of Gilmanton, and was incorporated in 1812. It is one of the most flourishing towns in the State. Two streams enter the lake from this town, namely, Gunstock and Miles Rivers.

Alton, on the south side of the lake, was incorporated a little more than seventy years since. It was formerly called New Durham Gore, and received its present name from one of the settlers, — being named for Alton in England. Merry Meeting or Alton Bay makes out from the lake about seven miles into this town, and receives Merry Meeting River. Mt. Major and Prospect Hill are the principal elevations; from the latter, the Atlantic is visible in clear weather.

The Cochecho River has its source just north of here, almost on the margin of the lake, and is a rapid and important stream; flowing south-east it is caught at Dover, and bound for the use and benefit of man. The Cochecho railroad has its terminus at Alton, and mountain travel from Boston, via Boston and Maine Railroad to Dover and the Cochecho to the lake, finds an easy and expeditious route to the magnificent scenery of Lake Winnipesaukee and its borders, as well as a short and convenient thoroughfare to the White and Franconia Mountains.

Smith's Pond, a collection of water some six miles long, is the source of a stream called Smith's River, which falls into the lake in Wolfboro'. There is a fine village on this stream overlooking the lake and surrounding mountains, affording a prospect grand in the extreme.

Copple Crown Mountain also affords a remarkable view, — the lake and its gems of islands; about thirty other lakes and ponds in New Hampshire and Maine; all the south-eastern part of the State; the grand hill to the west far up into the Pemigewasset country; and to the north the collection of hills as far as Mount Washington, whose grizzled summit towers up and overlooks the heads of all his fellows. All these combined are not excelled in picturesque and romantic grandeur.

There is a mineral spring in the town said to be similar and equal to the celebrated Saratoga, and is becoming a place of considerable resort. There is an excellent high school, and many other features important and attractive, and the trip among the beautiful and sublime scenery of New Hampshire has come to be regarded as incomplete, unless it includes Wolfboro'; and the crowning loveliness of this portion of the tour culminates in a moonlight excursion on the lake.

Tuftonborough, situated at the north-east extremity of the lake, was originally granted to John Tufton, grandson of John Mason, governor, who, in his will, made John and Robert Tufton heirs of his estates in New Hampshire, in consideration that they should assume the name of Mason; which requirement they complied with, and the name of John Tufton Mason is still legible in the old burying-ground near Christian Shore, in Portsmouth.

In several towns in this section of the State terms are used to designate many localities, which, though they may not be euphonious, are considered peculiarly appropriate and descriptive; for instance, a certain locality is known as "Barvel Whang," others as "Mackerel Corner," "Skunk's Misery," "Potatoborough," "Ossipee Pocket;" another is "Grasshopper's Grief," where it is said the soil is so destitute of vegetation that a pair of them would starve on a ten-acre lot without cut feed and meal at least once a day.

Fangs or lagoons of the lake extend into this town long distances in various directions, which are a paradise for pickerel, making this one of the best places for summer pickerel fishing in the State. Splendid views are obtained from many points in Tuftonborough, of scenery as wild, romantic and diversified as any in New Hampshire, though less extensive. From many of the eminences the prospect is charming; lagoons and estuaries of the great lake creeping quietly through narrow channels, hidden by luxuriant foliage, now reappear behind the intervening hills, and expand into broad sheets of water, or, following the natural depressions of the surface, take strange, fantastic shapes, presenting the appearance of an extensive system of ponds, with the great lake in the background stretching away to the south until lost to view among the numerous islands.

Meredith is on the west side of Lake Winnipesaukee. Meredith Village is beautifully and pleasantly located on a brook, which is the outlet of Goose or Measley Pond, which here falls into the lake. The soil is excellent, and few towns in the State excel it in the extent and variety of agricultural productions.

The road from Laconia to Centre Harbor runs principally through the town, and is a delightful drive; throughout the whole distance it follows the trend of the lake, which on the right stretches far away among the green islands until it is shut out from view by the luxuriance of their foliage; while on the other hand the neat farm-houses, the broad fields of corn and waving grain, and mowing-fields, the flocks and herds grazing on the green hill-sides, the great mountains towering up in all the surrounding background, the graceful steamers threading the mazy labyrinths among the isles, "walking the waters like a thing of life," while the gentle zephyrs lightly ripple the broad surface of the lake, — all combined in one grand sweep of the vision over the surrounding country, affords a scene which all can enjoy, but none describe.

Just here is the famous ahquedauken, or weirs, where the Indians took their food from the countless myriads of shad which crowded annually to and from the lake, the outlet, the Winnipesaukee River.

Dudley Leavitt, the celebrated "Old Farmer's Almanac" maker, lived and died in this town.

Sanbornton is a smart, thrifty town, situated on the Great Bay, or large body of water which receives and discharges the Winnipesaukee, or perhaps may be called an enlargement of that river. It is noted for the high character of its schools, for its general thrift, for its mechanical and manufacturing resources, and for its natural curiosities.

One of these is a gulf broad and deep, extending nearly a mile through hard, rocky earth. It is thirty-eight feet deep, and varies from eighty to one hundred feet in width, the walls corresponding so accurately as to present the appearance of having been rent asunder by some convulsion of nature. A cavern is also pointed out, extending horizontally some twenty-five or thirty feet into the earth, and is as likely a receptacle for Kidd's treasure or rattlesnakes as many others mentioned.

Near the head of Little Bay, on the Winnipesaukee, are the almost obliterated remains of an old fortress. It was evidently constructed by the Indians, and is a very ancient affair: relics of the aboriginal garrison have often been found in the vicinity.

It was believed that this was an important post on the great thoroughfare on the line of the trail from the Pennacook Confederacy along the lower Merrimack to Lake Winnipesaukee, branching eastward to the Pequauket country and the Amariscoggin, and westward by the head-waters of the Merrimack and the Connecticut to the St. Francis.

At the time the Sanborns and others came here to settle, the walls of this fortification were four or five feet high, and large forest-trees were growing within the enclosure, which gives to the ruins an air of antiquity. Pottery, tobacco, pipes, warlike and agricultural implements are among the trophies obtained by the ploughshare and the curiosity-seeker. This fortification is described as having consisted of six walls, one extending along the river and across a point of land into the bay, and the others at right angles, connected by a circular wall in the rear. Who shall say what terrible sieges this Malakoff of the wilderness has stood, what bombardments of arrows, what terrific attacks by water of some primitive Drake with his fleet of birch canoes, what heroic sorties with the death-dealing war-club and tomahawk, or the unconditional surrender, and the torture!

Gone, all gone, the barbarian actors in these rude conflicts, and records written in the sand or on the forest-tree, swept into the deep and wide grave of oblivion together, the few scanty records preserved having been rescued from the rubbish of uncertain tradition by the untiring, earnest, and generous efforts of the learned and thoughtful of another race of men, actuated perhaps more by a love of knowledge than of the race whose career they chronicle!

Sometime about 1746 or 1747, Col. Atkinson's force, near the outlet of the Winnipesaukee, built a strong fort for the protection of the exposed settlements on the Merrimack, and in the eastern part of the State. Some authorities have it that the fort at the head of Little Bay, which was generally supposed to be an undoubted relic of Indian engineering, skill, and labor, was constructed by this force while stationed there. The Winnipesaukee River passes through Northfield, having in it falls that afford the best of water-power, and are used for manufacturing, there being cotton and woollen mills and shops on them used for various purposes.

The New Hampshire Conference Seminary, an extensive and flourishing Methodist literary institution, is pleasantly located on a slight elevation, a short distance from, and overlooking, the river, and one eighth of a mile from Sanbornton Bridge. It possesses a valuable and extensive chemical and philosophical apparatus, and its collection of mineral specimens is large. This school is very pleasantly situated, its concomitants being a surrounding of charming scenery, a desirable quiet, economy of living, and a profitable and thorough system of instruction. Young gentlemen may here receive training of a high and enduring character.

The Winnipesaukee River, in its short career from the "weirs" in Meredith to its confluence with the Pemigewasset, just below Webster's Falls in Franklin, is an exceedingly ornamental stream, and useful for many purposes not enumerated, but especially so in its many mill privileges. The entire fall of this stream from the "weirs" to Franklin is nearly two hundred and fifty feet, two hundred of this being within four miles of its junction with the Pemigewasset in Franklin. When it is considered that the identity of the river proper is lost, for ten miles of its course, in the great bay lying between its source, the lake, and the forks of the Merrimack, it will be seen that it has no mean capacity for manufacturing purposes, and that this power is already quite extensively employed, with a still greater chance remaining to be some day more extensively operated.

Canterbury suffered severely from Indian depredations. So constantly did the enemy lurk about the town that the fields were cleared and tilled under the protection of a strong armed guard. In the year 1738, two men, named Shepard and Blanchard, were surprised by a party of seven Indians, who fired upon them, the first time without effect, when the two returned the fire, and Shepard escaped, but Blanchard was mortally wounded and captured, and died in a few days. They also captured a lad named Jackson, and a negro servant belonging to Thomas Clough, and carried them to Canada, where they remained till the close of the French and Indian war in 1749.

In 1752, two Indians, named Sabatis and Christi, came into the town, and, behaving friendly, were kindly treated, but after staying several weeks they suddenly left, taking with them, forcibly, two negroes. One of these escaped; the other was taken to Crown Point, and sold to an officer. The following year Sabatis returned to Canterbury with a companion called Plausawa. The former, on being remonstrated with for his past conduct, became exceedingly insolent, and, probably on account of being intoxicated, displayed a wild and threatening demeanor. They were followed and killed by some persons heedless of the injunction, "Vengeance is mine." These persons were arrested, ironed, and committed to jail; but the night previous to the trial a mob gathered, with crowbars, axes, and bludgeons, rescued them, and "Lo, the poor Indian" got no redress.

Shaker Village, built along the summit of a hill, is a compact, tidy, quiet, methodical community. The Shakers are a sect first known in this country just previous to the Revolution. It is believed to have been founded by Ann Lee, who came from England. The first society known in this country was established at New Lebanon, in New York, about 1780; and this one was organized two or three years later through the labors of two elders named Chauncy and Cooley.

The society occupy a large tract of land, nearly three thousand acres, which is owned and the labor performed in common. The buildings and the table are common. In short, most matters are common with them, while in others they regard themselves as marvellously proper. They consider that they are devoting their time, their toil, and their accumulations to the service of the Lord. Their work, farming, mechanical, and manufacturing, is always done in the proper season and manner, thorough, but in no sense wasteful. Their workshop and mills are models of neatness and convenience, peculiarly adapted by arrangement and fixtures to the kind of business carried on; their implements of the most improved pattern and material. The consequence of this thoroughness is, the Shaker products are sought at the highest prices. Agricultural implements, kitchen furniture, cloths, flannels, hose, herbs, and everything they have to sell find a ready market.

The laundry and dairy are arranged so conveniently and comfortably, that it seems as much pleasure to do the work as to see it done. A stationary engine does all the work of lifting, lowering, turning, washing, ironing, drying, churning, etc., and everything may be said to be done literally like clock-work, as they design each room shall be supplied with a clock, there being some hundred in the various rooms in the village. There is also a large bell, which calls the community to labor, meals, school, and devotion.

The Shaker barn is said to be the largest and best-arranged structure of the kind in the State. It holds from five to six hundred tons of hay, besides grain and other produce; two stables, each accommodating fifty cows, which are allowed to go in promiscuously, each knowing its name and stall, and all tied and untied by a slight and simple movement of a lever. There are also calf-pens, sheep-pens, places for sick animals, and the live stock of the Shakers here find a hotel and hospital combined. The barn is surmounted by a cupola, from which a good view of all the surrounding country is obtained.

They maintain a kitchen-garden of two or three acres, luxuriant with almost every variety of fruit and vegetables known to this latitude, with which their table in the season is bountifully supplied, they selling only what they cannot consume; and strangers are often supplied here with meals, always substantial and good, at a price never exceeding prime cost.

They do not believe in, or practise marriage at all, and regard Brigham Young with a feeling of abhorrence which language is inadequate to express; still instances have occurred where a brother and sister have come to an understanding, either by translating the silent but unmistakable language of love, or by surreptitiously obtaining an interview, and eloped, leaving their undivided interest in the society for their interest in each other.

The social theory and practice of the Shakers may be the correct one. Probably no one ever expressed or entertained a desire to interfere with their faith or practice; but there are those who regard it as simply absurd; while it is thought there are still many others who would remain unregenerated heathen, preferring to take their chance both in this world and the world to come, than be born again under the dispensation of Shaker Christianity without this dogma being expunged.

The Shakers are plain in language, manner, and dress, industrious, temperate, and healthy. Shaker Village seems a beehive where all are busy. Some one has said, — no doubt a slander, — that implements in the hands of the Shaker, whatever may be their design, are used to kill time; that his life is aimless; that he has no aspirations beyond his individual comfort; that he is destitute of the incentive to energy, enterprise, and ambition; in brief, that he is in the social order a nondescript; in religion a perpetual penance-doer; and in community a nonentity; in conversation "their yea is yea, and their nay nay." No one can dispute the fact that they are excellent citizens, conforming to all the requirements of citizenship, except military drill, — eschewing this as well as marriage as a part of their religion; and, taken as a whole, they are utilitarians enough to do as little harm in the world as possible, even though they may not do all the good they ought or might.

The surface of Pembroke is undulating. Its principal village, parallel with the Merrimack River, is five miles from Concord on an elevated ridge, and is between three and four miles long. Pembroke Academy, which has long been famous as one of the foremost educational institutions in the State, is on this street. A large tract of territory, embracing this town, was formerly known as Suncook, and was granted, under this name, by Massachusetts, in 1727, to Captain Lovewell and his band of Indian fighters, sixty in number, three-fourths of whom accompanied him in his last fatal expedition against Pequauket. On May 1st, 1748, James Carr, of this town, was killed by the Indians. Pembroke was incorporated under its present name in 1759.

The Suncook River is an important though not large tributary of the Merrimack. Its source is a pond on the summit of one of the Suncook mountains, — a chain of high hills located in Gilmanton, and running north and south through that town. This pond is located about one thousand feet above the base of the mountain, and its outlet, on descending, falls into another pond; from thence into still another, from which it emerges, and receives, in its course, many tributaries. This river is very rapid, affording numerous fine mill privileges along its whole course.

Gilmanton, where the source of the river is located, also gives rise to the Suncook, and beside these is watered by the Winnipesaukee. This was formerly a very large township, extending northward to the shores of the Winnipesaukee, but has, from time to time, been shorn of its extensive proportions, the towns of Guilford and Upper Gilmanton having been carved out of its original territory. Iron ore was at one time obtained in considerable quantities, and iron works were established.

Academy Village is a very beautiful and unusually pleasant place. The soil is very fertile, and in a high state of cultivation; the dwellings spacious and neat; the people enterprising, thrifty, and intelligent, — the facilities afforded by the Gilmanton Academy being ample.

"Porcupine Hill is a remarkably abrupt precipice of granite, gneiss, and mica slate rock, which form, by their overhanging strata and deep ravines, a pleasant and favorite resort of the students of Gilmanton Academy, — an old and highly respectable institution of learning. Below this steep precipice is a deep and shady dell, thickly clad with a dark evergreen foliage of forest-trees, while the rocks are wreathed in rich profusion by curious and beautiful lichens or mosses. Wild plants are abundant and various.

"Gilmanton Academy was incorporated October 13, 1762." *

The town was granted, in 1727, to one hundred and seventy-six persons, twenty-four of whom were named Gilman, but the settlement was delayed for many years on account of the hostility of the Indians. In 1761, Benjamin and John Mudgett, with their families, settled here, and Dorothy Weed was the first white child born

---

* "New Hampshire As It Is."

in Gilmanton. Hon. William Badger,* formerly Governor of New Hampshire, was a native and resident of this place.

The Suncook River takes a south-east course through Gilmanton Village, passing into Barnstead, which is a thriving town, almost wholly engaged in agricultural pursuits. The township was granted, in 1727, to Rev. Joseph Adams and others, but its settlement was delayed nearly forty years. It lies in gentle, grand, and sometimes abrupt undulations, and is dotted with several fine ponds, which discharge their waters into the Suncook River. The largest of these are the two Suncooks, near together, Half Moon and Brindle.

The variety of minerals yet discovered is not extensive. Yellow ochre, iron, plumbago, and specimens of basaltic trap-rock are among the number.

Barnstead was incorporated in 1767. In 1807, a very useful institution was organized and incorporated, called the "Social Library," which is still in a flourishing condition, and is a source of great benefit to the citizens, who are unusually intelligent and well-informed. From this town, the Suncook, much increased in volume, passes into Pittsfield, where it furnishes many excellent mill privileges.

Pittsfield, though not yet a century old, is a populous and flourishing town, which is mainly due to the water-power of the Suncook, the surface of the town being mountainous and broken. Pittsfield Village is located in a deep valley, and is surrounded by high wooded hills, — a picturesque and romantic scene. The village is large and lively, surpassing most inland towns destitute of railroad facilities. Activity and business are its concomitants, and when the railroad, now in contemplation, is completed to Manchester, furnishing easy and rapid communication with the outside world, Pittsfield will at once become an important town. It has a cotton manufactory, employing nearly two hundred hands, with a capital of two hundred thousand dollars.

Catamount Mountain is situated in Pittsfield, and is fifteen hundred feet high. The view from its summit is extensive and grand;

---

* Gov. Badger's widow is still living. She resides in Upper Gilmanton, and, though seventy-eight years of age, traces of those attractions of mind and person, which distinguished the noble dames of a past ago, still exist.

the broad Atlantic is plainly seen, as well as islands, headlands, and the great circle of mountains, extending from Agamenticus in Maine, to the White Hills on the north, and the high summits on the west, including the grand Monadnock. There is a fine sheet of water on the mountain, which is about one hundred and seventy-five rods long, fed by springs. There are other ponds in the town, and from Wild Goose Pond bog iron ore has been taken in large quantities. There is also a mineral spring, the impregnation being a compound of sulphur and chalybeate. There are many peat meadows, some of which have been reclaimed, and are remarkably productive. In some parts of the town the magnetic needle shows considerable variation, which is attributed to the proximity of minerals in large masses; black tourmaline and magnetic iron ore have been found in several places. Tourmaline is usually found in hexagonal or triangular prisms, terminated by three-sided pyramids, — black, but sometimes green, blue, red, and brown. It was formally known under the name of schorl; and its crystals, when heated, are remarkable for exhibiting electric polarity, — attracting and repelling the needle.

There is a society of Friends in Pittsfield, who have a meeting-house, and maintain regular public religious worship.

Epsom was granted, in 1727, to Theodore Atkinson and others, and, like many other towns whose names are not of Indian origin, was named for a place of the same name in England. The town is well watered by great and little Suncook Rivers and many smaller streams. There are also three ponds in the town.

McKoy is the highest elevation, and derives its name from Mrs. McKoy, who was taken captive by the Indians in 1747, and did not return till after the close of the French and Indian War.

The soil of Epsom contains several minerals, some of them rare; brown oxide, sulphuret of iron (copperas), terra sienna, mineral paint, — an alluvial deposit, — arsenical pyrites, silver mixed, or argentiferous galena, and hydrated oxide of iron, mixed with crystallized quartz.

On receiving the news of the battles of Lexington and Concord, twelve hundred of the sturdy yeomanry of New Hampshire hastened to Boston. Among them was Major McClary, of Epsom, who fell in the battle of Bunker Hill, while gallantly fighting for his country.

Allenstown, on the Suncook River, was granted in 1731. In 1748, as Mr. Buntin, his son, and James Carr were at work on the west side of the Merrimack, opposite the mouth of the Suncook, they were set upon by Indians, and Carr, attempting to escape, was shot; Buntin and his son were made captives, taken to Canada, and sold, but made their escape, and returned home in about a year.

Allenstown was not incorporated until one hundred years after it was granted. Bear Brook is a large stream, and furnishes many mill sites, and is one of the most famous trout brooks in this section of the State, furnishing good fishing-ground — with its tributaries — some thirty miles in extent. Suncook River divides this town from Pembroke, and has a splendid water-power at the village, which takes the name of the river not far from its confluence with the Merrimack. It seems to be a well-established fact, that clearing the lands has diminished the size of the rivers, and that those in New Hampshire are small now compared to what they were previous to its settlement. A proof of this may be deduced from an extract of the address of Gov. Wentworth to the Legislature in 1746.

"After the mischief was done by the Indians at New Hopkinton, the inhabitants of Canterbury were in the utmost distress, for a great number of the inhabitants, . . . then in the woods, which occasioned an alarm in that quarter; and being apprehensive the enemy had besieged that garrison, I ordered a detachment of Capt. Odlin's and Capt. Hanson's horse to march out to their relief. For want of a bridge on Suncook River, both detachments were obliged to march more than double the distance, and as Canterbury is the only magazine for provision on our frontiers, I hope you will think it worth your consideration, that a bridge be built here as soon as the weather will admit of it."

It will be seen that a century since, when the country of the Suncook and its branches was an unbroken wilderness, cavalry troops were forced to make a wide circuit towards its source to find a fording-place. Now, however, there is scarcely a point in its whole course where an athletic youth of fifteen could not ford it, except at times of high freshets.

The mills on the Suncook, a short distance from its confluence with the Merrimack, obtain their motive power from the former stream. The fall, at this place, is picturesque, romantic, and grand; the height being ample; but in dry seasons the volume of water becomes much attenuated. There are three corporations at this place, known as the Pembroke, capital five hundred thousand dollars; Webster, capital five hundred thousand dollars; China, capital one million dollars. Garvin's Falls, a short distance above Suncook, on the Merrimack, are considered available for manufacturing purposes, having a perpendicular fall of twenty-eight feet, fully equal to that at Lawrence, and the time may come when a flourishing manufacturing city will cluster around this waterfall. ' Should this be the case, and should Sewall's Falls, in the north part of Concord, be improved. the result would be an unbroken city, stretching along the Merrimack, on both sides, from Fisherville on the Contocook, to the Suncook, a distance of fifteen miles, and even to Hooksett, a smart manufacturing village two miles farther down the Merrimack.

The Contoocook River, a very considerable tributary of the Merrimack, has its principal sources on the height of land between the waters of the Connecticut and Merrimack Rivers in Rindge and Jaffrey, Cheshire County. In the town of Dublin, which is also located on this elevated ridge, one of the churches sheds the rain on one side into the Ashuelot, and thence into the Connecticut, while the other side is drained through the Contoocook into the Merrimack.

The Contoocook has its course through a portion of three counties, and is a very important stream, whether we consider the large tract of territory watered by itself and numerous affluents, the great amount of excellent farming lands which border it, or the many mill privileges, some of them very superior ones, it furnishes along the line of the main stream and its branches. It flows in a general north-east direction, and, passing out of Hillsboro', enters Merrimack County at Henniker, pursuing its way through several important towns, and receiving the valuable accession of Warner and Blackwater rivers in Hopkinton. It now becomes a large stream, and falls into the Merrimack at the south-east angle of the town of Boscawen, being the division line between Concord and the above-named town.

Hillsboro' has a very uneven surface; the soil, however, is generally strong and productive. It was originally settled, in 1741, by James McCalley and others, and was called township number seven. The wife of McCalley was the only woman in the settlement the first year. Three years after this settlement was made, it was abandoned on account of the French War, and was afterwards granted by the Masonian proprietors to Colonel John Hill, of Boston, who, with others, settled it in 1757, and from him received its present name, being incorporated in 1772. The principal pond, which is not large, bears the name of one of the original settlers, and is called Lyon's Pond.

This town is amply watered by Contoocook and Hillsboro' Rivers, and many other smaller streams, furnishing mill-sites with good power, which are improved, several kinds of mechanical works being carried on, as well as manufacturing in a small way. The largest is a cotton-mill, which has sometimes employed as many as twenty-five or thirty persons. Veins of plumbago of remarkable purity have been found and worked to a considerable extent. Hillsboro' is in many respects a noted town. For intelligence, integrity, industry, and sobriety it is believed to be unsurpassed, while thrift and general content, certain to be evolved from this favorable condition of society, are seen on every hand.

Hillsboro' was the adopted residence of Benjamin Pierce, who was born in Chelmsford, Massachusetts, in 1757. He was the son of a farmer, and lived in an age when education was not only not generally diffused, but was difficult to obtain anywhere; and in a land so new and sparsely settled the facilities for learning were slender indeed.

The news of the battle of Lexington was conveyed to him, as it was to others, "by the man on horseback;" and, without hesitation or delay, he took his gun and started immediately on foot and alone for Lexington. Finding the British troops retreating on Boston, he fell in with the Provincials, and pursued them as far as Cambridge, where he enlisted for the war, and was in the memorable battle of Bunker Hill, and in active service until the close of the Revolutionary War, having risen, by promotion for gallant and meritorious conduct, from a private to a captaincy. On returning to peaceful pursuits, after nine years of arduous service in his country's cause, he found the Continental currency so depreciated that the total amount of his pay was not sufficient to buy a farm; consequently he again took up the line of march for a wilderness country where land was so cheap as to be within the reach of his slender means, and located in Hillsboro'.

Endowed by nature with superior intellectual abilities, he had, by perseverance in watching and appropriating every opportunity, overcome the imperfections of his education, and was at once recognized as one of its ablest and most intelligent members by the community in which he lived. He did not, however, retire from the military service, but continued to serve in the militia for many years, being at one time colonel of the famous regiment which furnished the gallant McNeil and Miller for the war of 1812, and was a general in the State militia at the time he retired from the service. He held many civil offices, and in 1818 was elected sheriff of the county. Finding several poor debtors in Amherst jail, one of them having been incarcerated four years for no crime, save that he owed a debt which he could not pay, kind-hearted and generous, he at once paid their debts and liberated them. Addressing them briefly, he said : —

"I have a duty to perform. I must either be governed by the law, and suffer you still to remain the devoted victims of unavoidable misfortune and honest poverty, shut out from the genial light of heaven and the vital air, God's equal gift to all, to endure, perhaps perish under, the privation incident to your situation and the stern ravages of approaching winter, forlorn and destitute, with no friend to comfort. no society to cheer, no companion to console you; or I must be directed by the powerful impulse of humanity, pay the debt myself, and bid you leave this dreary and gloomy abode. My unfortunate fellow-citizens, my duty to myself will not suffer longer to remain here an old companion in arms, who fought for the liberty of which he is deprived, for no crime but that of being poor. . .

"In this view, go; receive the uncontaminated air which is diffused abroad for the comfort of man; go to your families and friends, if you have any. Be correct in your habits. Be industrious; and if your tottering and emaciated frames are so far exhausted as to prevent your getting comfortable support, apply to the good people for relief. And may the best of Heaven's blessings accompany you the remainder of your days."

General Pierce was elected to the principal offices in the State, which he filled with credit to himself and advantage to the people. He was governor in 1827, and again in 1829. Governor Pierce was, even to his last years, remarkably social and genial, and was, consequently, a companion for the grave and gay alike, his society being sought by the youthful and vivacious as well as the sedate and thoughtful.

This distinguished man, who possessed a lofty and resolute intellectual, and rugged physical organization, was still further distinguished by being the father of one of the Presidents of the United States.

"On the 26th of December, 1825, it being his sixty-seventh birthday, Gen. Benjamin Pierce prepared a festival for his comrades in arms, the survivors of the Revolution; eighteen of them, all inhabitants of Hillsboro', assembled at his house. The ages of these veterans ranged from fifty-nine up to the patriarchal venerableness of nearly ninety. They spent the day in festivity, in calling up reminiscences of the great men whom they had known, and the great deeds they had helped to do, and in reviving the old sentiments of the era of seventy-six. At nightfall, after a manly and pathetic farewell from their host, they separated, 'prepared,' as the old general expressed it, 'at the first tap of the shrouded drum, to move and join their beloved Washington and the rest of their comrades who fought and bled at their sides.' " *

He died in 1839, at the mansion in Hillsboro', which had succeeded the log cabin, on the very spot, now blooming and adorned, where, fifty years before, his sturdy blows first made a clearing.

The late Hon. Chandler E. Potter was for several years a resident of Hillsboro'. Judge Potter was for many years justice of the Police Court of the city of Manchester, and was, perhaps, more thoroughly versed in Indian lore and history than any other man in New England, if not in the country. His admirable "History of Manchester" exhibits a vast research in this, as well as in history proper, and in his unexpected decease the community has been deprived of a historical repository as extensive and varied as interesting and useful.

Henniker is an uneven but very excellent farming town. The hilly lands are unusually productive in the way of cereals and grazing facilities; the valleys, meadows, and intervals yielding well in

* Hawthorne.

hay, corn, and common vegetable products. The town was granted by the Masonian proprietors, under the name of "Number Six," to James and Robert Wallace and others, in 1752. The first house, constructed of logs, was erected in 1761 by James Peters, who was the original settler. It was incorporated in 1768, receiving its present name from John Henniker, Esq., a London merchant, who was a friend of Governor Wentworth and a member of Parliament.

The Contoocook River, which in its course through several towns is so winding or crooked that it is said to "run more than three times its length," passes through the centre of the town. Besides machine shops there are mills where several kinds of woollen fabrics are produced. Craney Hill is the highest elevation of land in the town. On the summit of this hill is a large boulder, some fifteen feet in diameter, an undoubted deposit of the ancient glacial drift, released from its icy embrace by contact with the obtrusive crown of this hill, at the time the earth was being evolved from a submarine condition, by an application of the principle of hydrostatic engineering power which has never since been equalled. This boulder is so poised, that, only for its having been riven (supposed by lightning) and divided into two nearly equal parts, half a dozen men could move it from its present position, and it could not stop until it had rolled a distance of more than three miles. There are in the country several cases of the transplanting of hill-tops and mountain-peaks,* and the subject is attracting much attention from scientific men, as the "'Testimony of the Rocks" is regarded as incontrovertible. The Contoocook Valley Railroad, extending from Hillsboro' to Concord, passes through Henniker, affording good facilities for freight and travel.

Bradford, which is the northern boundary of Henniker, a little

---

* Prof. Gunning delivered a lecture in Hartford, Conn., recently, on the last glacial period, during which he stated that he had seen in Stamford, Vt., a mountain of granite as peculiar as that of Superior, but of different type. The crystals were foliated. Science can find that granite at home only in Stamford. The mountain is a truncated cone. The top has been clipped off. North of the mountain there was not a single boulder of foliated granite. South of the mountain there were multitudes of such boulders. Perched on the very top of Hoosac Mountain the tourist may see a boulder about seventy feet in circumference and fifteen feet high. If he looks at the boulder, then at the mountain, he will see that the boulder has no kinship with the mountain. The boulder is that same Stamford granite, — a Vermont "carpet-bagger" ensconced on one of the highest peaks of Massachusetts. The tourist may look south-westward over Deerfield Valley, thirteen hundred feet deep, and see far in the distance the outlines of Stamford Mountain, from whose top that boulder was torn.

to the west of the Contoocook, is equidistant from the Merrimack and Connecticut Rivers, and contains a very singular natural curiosity. Todd's Pond, which extends into Newbury, is the largest collection of water in the town, and has several floating islands. Bradford Pond, containing a large number of islands, and noted for its romantic surroundings, is justly celebrated for its wild and picturesque scenery. The waters discharged from these ponds find their way into the Contoocook.

Bradford was settled in 1771 by Dea. William Presbury, and was incorporated in 1787, including in its limits a portion of the territory of the present town of Washington. Several years since a stone quarry of good quality and great value was discovered here, which has been extensively wrought.

Washington, the western boundary of Bradford, is one of the most remarkable towns, in some respects, in the State. It was originally granted by the Masonian proprietors to Reuben Kidder, Esq., who settled it in 1768, under the name of Camden. It was incorporated Independence year (1776) by the distinguished and illustrious name of the Father of his country. Washington is principally an agricultural community, the exception being — as timber is of superior quality, plenty, and cheap — the manufacture of various kinds of wooden ware. There are more than twenty ponds in the town, and, of course, a very large number of brooks and small streams. These waters are very pure and well stocked with the best kinds of fish.

Lovewell's Mountain, an elevation of historic interest, is situated in Washington, and derives its name from Captain Lovewell, the famous Indian hunter; from its summit an extensive view is obtained of the surrounding country, and from this observatory he was accustomed to scan the forest around, to ascertain if there were any savages lurking near; if there were, the smoke of their wigwam fires could be seen curling above the tree-tops. On one occasion, as he was splitting logs near the summit, he was suddenly surrounded by seven Indians. A resort to force, unarmed and defenceless as he was, against seven stalwart savages, was entirely out of the question; and, determined to escape, stratagem was his only alternative. Consenting to go with them into captivity without hesitation, he requested them as a simple favor to assist him to split the log on which he was at work when they interrupted him. They consented without suspicion, and, ranging themselves on either side of the log, they put their hands into the opening where it was partially riven by the entering wedge, to assist with all their force in pulling it asunder, when Lovewell suddenly striking out the wedge, the fissure closed up, and they were all securely caught in this novel man-trap. Lovewell then despatched them at his leisure, scalped them scientifically, and continued his labor.

Besides the liberal provision for common schools, Tubbs' Union Academy, a flourishing educational institution, supplies its pupils with a high order of instruction.

Warner is the next town west of Boscawen, and lies immediately at the base of Kearsarge Mountain. It was granted by Massachusetts, to Deacon Thomas Stevens and sixty-two others, in 1735, under the name of "Number One." It was afterwards called New Amesbury, and was subsequently regranted by the Masonian proprietors, but a controversy arose between the latter grantees and the earlier claimants, which was not adjusted for several years, and the town was not incorporated until 1774, when it assumed its present name.

Warner was settled, in 1762, by David Annis and Reuben Kimball. Warner River, a tributary of the Contoocook, rises in the Sunapee Mountains, and passing through this town affords many good privileges for mills and shops. Pleasant Pond, whose waters are cold, pure, and deep, has no visible inlet or outlet, although it maintains its maximum depth in the dryest season. The soil of this town is very good: the minerals are gneiss and mica slate, the latter containing beds of limestone and talcose rock, while in the former are found very fine specimens of splendidly colored pyrope, a garnet sometimes blood-red, but generally a modification of that color, and frequently tinged with yellow. There is an extensive and valuable soapstone quarry, and also a large and rich peat bog, which is twenty-five to thirty feet in depth, from which beaver cuttings have been exhumed, denoting that in the far-distant past it was the site of an immense beaver dam.

Hopkinton joins Concord on the west, its principal village being about seven miles from the State capital. It was granted by Massachusetts, in 1735, to John Jones and others, and was called "Number Five," and was subsequently called New Hopkinton, from Hopkinton in Massachusetts. It was originally settled in 1740, but, on account of the French and Indian war, the settlement was abandoned until after its close. In 1746, the Indians surprised the garrison, and captured eight persons and carried them away. In 1753, Abraham Kimball, the first white male person born in the town. and Samuel Putney, were captured by the Indians. Three days afterward the Indians were surprised and attacked in Boscawen, and Putney was recaptured, while Kimball made his escape by the aid of a sagacious dog, which ferociously attacked an Indian as he was about to bury his tomahawk in Kimball's skull. In 1756, Henry Miller and others received a grant of this township, which occasioned an acrimonious controversy, lasting several years, which was finally adjusted and peace restored by an act of incorporation in 1765, under the name of Hopkinton, — the "New" having evidently been worn off, or at least left off.

The Contoocook River meanders through Hopkinton, receiving in its passage the Warner and Blackwater, and furnishes an excellent water-power. Contoocookville, located at the junction of the Merrimack and Connecticut Rivers and Contoocook Valley railroads, is a lively village, and a place of considerable business. In the village are shops and mills, and large quantities of lumber are manufactured, as well as in other parts of the town, and shipped to distant markets. The intervals along the rivers are exceedingly fertile, while on the undulations are seen many tracts of highly cultivated and productive lands. The people are generally engaged in farming, to which the land is mostly well adapted, and many varieties of delicious fruit are grown, the soil being unusually good for that purpose, and more than ordinary attention is paid to its culture.

CHAPTER VII.

Concord. — The Pennacooks. — First Settlement. — State Institutions. — Ex-President
Pierce. — Isaac Hill. — Count Rumford. — Bow. — Hooksett.

HAVING explored the head-waters of the Merrimack, and entered the vast amphitheatre where it first appears to a very select audience, and where from tiers of high encircling benches the gods of the mountains look down upon its opening career, its "leap for life," its noisy tumbling from rock to rock in sheets of milky foam, as it were, into real, tangible being; where they from their thrones rich draped in fleecy clouds, and the invisible subjects who people their dreary realms, almost "solitary and alone," look down upon its slender proportions, listen to its liquid, soul-inspiring music, and bid it God-speed on its journey of usefulness and toil; following it on this journey along the route it has chosen to take, and traversing the course of each of its principal branches for the purpose of obtaining desirable and reliable information, matters of fact, of importance, and of interest, and also to procure such information as will most clearly disclose and exhibit a modicum of the importance and value the Merrimack River already is, as well as the undeveloped power it still possesses to the people of New Hampshire and New England, and leaving the principal forks, a short and pleasant march of twenty miles, and the beautiful city of Concord is reached.

Concord is the heart of the Commonwealth, the Merrimack River the main artery, and its tributaries the veins and capillaries through which its throbs and pulsations are felt in the most remote extremities of the State. The history of Concord, anterior to its occupation by the pale-face, as well as during its progress to its present condition of importance in the way of wealth, population, power, and influence, seems to have been intimately connected with that of the fine river which has its course through the centre of the State, and this, its capital city. It has been the fortune of Concord, whether good or otherwise, to be the chosen seat of government of more than one race of men. When the Pennacook nation of confederate tribes was in the zenith of its power and glory, and its jurisdiction and influence extended beyond the present limits of New Hampshire; when the Pennacook made laws and made wars, and made peace by conquest or treaty; fixed the legal standard of his shell circulating medium; devised, like his more civilized successor, oppressive "internal revenue" laws; replenishing his depleted exchequer as occasion required for the maintenance of the dignity and efficiency of his government, by the collection of game, fish, implements, and wampum, and did those things which a sovereign power may of right do, — he also had the seat of his government, and the visible awe-inspiring emblems of authority, and the head-quarters of his foremost sagamon at Concord.

"When the red man used this river, as the white man now does, for important purposes in the economy of life, as a means of transportation and communication, cultivating, after his rude and primitive fashions, its fertile intervals for a meagre supply of maize and vines, and procuring a material portion of his sustenance from its waters, here at Concord was the recognized and legal seat and centre of his dominion, and he called it Pennacook from pennaqui (crooked), and auke (place), the river winding somewhat circuitously through this section, and embracing the great intervals of Concord within its folds; or, it may be from penak (a ground nut), and auke (a place)." *

The Pennacooks were the most important and powerful tribe of Indians on the Merrimack River, from the fact, no doubt, that their location was far superior to that of any other tribe, a longer established and more permanent resident, and above all, the extraordinary ability and wisdom of their great sagamon, Passaconaway.

Their acknowledged superiority naturally placed them at the head of the powerful confederacy of the Merrimack River tribes and the Winnipesaukees. Amoskeag, Souhegan, and Nashuas acknowledged their sway, and were tributary to them; the Wamesits, naturally inferior in many respects, and being intermarried, were also tributary, and ultimately, the several tribes along the Merrimack ceased to maintain an independent existence in name as well as in fact, and

* Potter.

were absorbed by the Pennacooks. The names of the tribes so merged were the Agawams, located in what is now Essex County, Massachusetts, Wamesits, Nashua, Souhegan, Namoskeag, and Winnipesaukee. Besides these, the following tribes, not on the Merrimack, acknowledged fealty, were tributary to, and confederates of the Pennacooks: the Wachusetts, Coosucks, Pequakuakes, Ossipee, Squamscotts, Winnecowetts, Piscataquaukes, Newichewannocks, Sacos, Amariscoggins. As it is interesting to know the derivation and significance of the names by which these aboriginal communities were known and distinguished, some of them are here given.

"The Winnipesaukees occupied the lands in the vicinity of the lake of that name, one of their noted fishing-places being at the outlet of the Winnipesaukee, now known as the weirs, — having remained at that place long after the advent of the whites. Winnipesaukee is derived from winne (beautiful), nipe (water), kees (high), and auke (a place), meaning literally the beautiful water of the high place. Wachusetts from wadchu (a mountain), and auke (a place), near Wachusetts Mountain in Massachusetts; the Coosucks, from cooash (pines), upon the sources of the Connecticut River; the Pequaquaukes from pequaquis (crooked), and auke (a place), upon the sources of the Saco, in Carroll County, in New Hampshire, and Oxford County, in Maine; the Ossipees from cooash (pines), and nipe (a river), upon the Ossipee Lake, and river in Carroll County, New Hampshire, and York County in Maine; the Squamscotts, from winne (beautiful), asquam (water), and auke (a place), upon Exeter River, in Exeter, and Stratham in Rockingham County; the Winnecowetts, from winne (beautiful), cooash (pines), and auke (a place), in the same county; the Piscataquaukes, from pos (great), attuck (a deer), and auke (a place), upon the Piscataqua River, the boundary between New Hampshire and Maine; the Newichewannocks, from me (my), week (a contraction of weekwam, a house), and ouanniocks (come), upon one of the upper branches of the same river; the Sacos, from sawa (burnt), coo (pine), and auke (a place), upon the Saco River in York County, Maine; and the Amariscoggins, from namaos (fish), kees (high), and auke (a place), upon the Amariscoggin River, having its source in New Hampshire, and emptying its waters into the Kennebec." *

* Potter.

All the tribes of the interior were known to the Indians living near the salt water, as Nipmucks, or fresh-water Indians, that being the general name by which all were designated. "Nipmuck is derived from nipe (still water), and auke (a place), with the letter M thrown in for the sake of the euphony;" and, true to their name, the Nipmucks usually had their residences upon places of still water, the ponds, lakes, and rivers.

This was the great Pennacook Confederacy; having its settlements scattered over a vast extent of country, its communication kept open by lines of fleet runners, it was powerful for mischief, and whether its ire was aroused by its implacable enemy, the Mohawks, or sought to wreak vengeance on the encroaching colonies, it concentrated its forces, and moved suddenly and rapidly, creating alarm, terror, and dread. Pennacook was fortified for the protection of habitations, utensils, supplies, and non-combatants, as well as a safe retreat from the war-path when a foe was encountered, whose force was either too strong or unknown; and when the danger was past, or the alarm was discovered to be false, he could easily replenish his exhausted stores by following the Merrimack up or down, or the Contoocook on the west, which, as also its tributaries, was stored with fish, and its forests abounding in game, while on the east, the Soucook and the Suncook were famous fishing-places, and deer and other game filled the woods; the Dark Plains of Pennacook, Pembroke, and Chichester being a rendezvous where he could assemble for the chase, secure and unobserved, and without alarming the game, and range this famous deer country to Catamount Mountain and to Deerfield, without danger of interruption until at least his present wants were amply provided for.

The Pennacook, doubtless, had the sagacity to see at an early period after the advent of the pale-face, his ultimate and inevitable doom, and being human, though an Indian, consequently possessed a spirit not altogether praiseworthy, which was no doubt aggravated by the vicious blood of his race, and it may not be wondered at that he determined to leave the mark of his resentment and his vengeance whenever and wherever he could. On the other hand, however laudable the motives by which he was actuated, the white man was the despoiler of his home, his patrimony and his hopes, to him only an interloper and a robber; he came upon the Indians' domain, the title to which had remained unquestioned and undisputed for ages, and while the intruder bore the Christian badge and banner, on which was inscribed " Peace on earth, and good will to all men," and the golden rule, he rudely thrust the simple child of Nature from the soil of his nativity, and where he did not choose to " stand not on the order of going, but go at once," but insisted with some show of reason on his right to remain, the pious pilgrim was forced to kill him to get rid of him, while at the same time he gave glory to God that he had received the light, and was not a pagan savage.

Concord was first visited by white men as early as 1639, but for half a century it remained too far beyond the frontier, and was consequently too warm with Indian atrocities, actual or threatened, for the prosperity, comfort or health of civilized people. In 1725, it was granted to Benjamin Stevens, Ebenezer Stevens, and others, by Massachusetts, which had usurped jurisdiction over a very large portion of the territory of New Hampshire, under the name of Pennacook Plantation. The next year something more than a hundred building-lots were laid out along the Merrimack, and some fifty or sixty people commenced building and farming operations. A house for public worship on the Sabbath, and defensive works for the protection and security of the settlers were also commenced, but were not completed until the following year. At the same time a house for the Rev. Mr. Walker was erected, which is still standing, or was quite recently, and is believed to be the oldest two-story house in the State. Edward Abbott's house was built at the same time, and the first two children, a girl and a boy, of the Caucasian race, were born in this house. The former died in 1797, and the latter in 1801. This building is still standing on Montgomery Street, but is now used for a barn.

The first town-meeting of Pennacook Plantation was held January 11, 1732, and Captain Ebenezer Eastman was elected moderator; and this was probably the last one also; for the next year the General Court of Massachusetts passed an act of incorporation, and changed its name to Rumford. Thirty years later, by order of the king in council, Rumford was declared to be within the jurisdiction of New Hampshire, and, in 1765, it was incorporated by the latter power under its present name.

During all these years the Indians remained in unpleasant proximity, and maintained a menacing and hostile attitude, making frequent predatory incursions, rendering the settlers insecure in their lives as well as effects. In 1739, for better securing themselves against an apprehended attack of the Indians in strong force, a garrison was erected. This fort was large, including the dwelling-house of the Rev. Mr. Walker, and designed as a refuge for the defenceless in case of emergency.

In 1742, the wife of Jonathan Eastman was taken captive and carried to Canada, but was subsequently ransomed and returned to her friends. On Sunday, August 10th, 1746, a band of savages, supposed to number one hundred or more, secreted themselves near the meeting-house, intending, doubtless, to attack the people while at worship; but fortunately a scouting party of between thirty and forty men, under command of Captain Daniel Ladd, of Exeter, opportunely arrived on that day; besides this, the men went to church well armed, and thus the murderous designs of the enemy were frustrated, and, instead of finding the settlers an easy prey, and making them their victims, they were attacked and driven away. Chagrined at their discomfiture, the Indians secreted themselves in the woods, about a mile from the village, and near the path which led to Eastman's Fort, in the west part of the town, for the purpose of intercepting or ambuscading squads of soldiers, supposing no doubt that they would pass to and from the settlement and fort. This scheme was well considered, the position well chosen, and the success terribly complete. In the morning, Lieut. Jonathan Bradley, of Capt. Ladd's company, and seven men started for Eastman's Fort, and having proceeded a mile and a half, were attacked by the Indians, and what followed may be told by Abner Clough, of Nottingham, clerk of the company, who kept a record of its doings:

"And when they had gone about a mile and a half, they were shot upon by thirty or forty Indians, if not more, as it was supposed, and killed down dead. Lieut. Jonathan Bradley and Samuel Bradley, John Lufkin, and John Bean, and this Obadiah Peters, — these five men were killed down dead on the spot, and the most of the men were stripped; two were stripped stark naked, and were very much cut and stabbed and disfigured, and Sergeant Alexander Robberts and William Stickney were taken captive, and have never been heard of since. It was supposed there was an Indian killed where they had the fight; for this Daniel Gilman, who made his escape, saith that he was about sixty rods before these men when they were shot upon, and he says the Indians shot three guns first. He says that he thought our men shot at a deer: he says that he run back about forty rods upon a hill, so that he could see over upon the other hill, where the Indians lie, and shot upon the men, and he says, as ever he came upon the hill so as to see over upon the other hill, he heard Lieut. Jonathan Bradley speak and say, 'Lord, have mercy upon me; fight.' In a moment his gun went off, and three more guns of our men's were shot, and then the Indians rose up and shot a volley, and run out in the path, and making all sorts of howling and yelling; and he did not stay long to see it, he said. It was supposed that John Lufkin and Peters were the first shot, as they were in the path about twelve or fourteen rods apart, and they shot Samuel Bradley, as he was about twelve feet before where this Obadiah Peters lay, and wounded him so that the blood started every step he took. He went about five rods right in the path, and then they shot him right through his powder-horn as it hung by his side, and so through his body, and there lay these three men lying in the path, and Lieut. Bradley run out of the path about two rods, right in amongst the Indians. He was shot through his waist; it was supposed he killed the Indian. It was supposed that he fought (as he stood there in the spot where he was killed) till the Indians cut his head almost to pieces, and John Bean run about six rods out of the path on the other side of the way, and then was shot right through his body, so that there was none of these men that went one or two steps after they were shot, excepting this Samuel Bradley, that was shot as above said, and there seemed to be as much blood where the Indian was shot as there was where any one of our men were killed. It was supposed the men lie there about two hours after they were killed before anybody came there. We did not go till there came a post down from the fort, about three quarters of a mile beyond where the men lie, and were killed."

A granite monument has been erected on the spot, to commemorate this bloody event, by Richard Bradley, Esq., a grandson of Samuel Bradley, one of the slain.

In 1808, the capital of the State was established permanently at Concord, and, in 1816, the erection of the State House was commenced, and was ready for occupation three years afterwards. The

State House grounds comprise two acres extending from Main to State Street, substantially enclosed and ornamented with beautiful shade-trees. The entire original cost of land and building was a little more than eighty thousand dollars, but alterations, additions, and improvements have been made, which will probably sum up a total cost of not less than two hundred thousand dollars.

In 1823, the County of Merrimack was erected by taking a number of towns from Rockingham County on the east side of the river, and from Hillsboro' on the west, in honor of the beautiful river passing through it. Concord being much the largest town, in population as well as area, was selected at the same time as the county-seat, and is provided with suitable county buildings. It contains an area of forty-two thousand acres, one acre in every twenty being covered with water. Much of the land is meadow, or interval, composed of rich alluvial and vegetable deposit, which formerly afforded the Pennacook ample breadth for his scanty husbandry, and where he could produce material for his nasamp and succotash; and it is worthy of note that the uninterrupted cultivation of these lands for two centuries has scarcely impaired their fertility or diminished their productiveness.

The county court-house, built in 1855, is a very fine building, and took the place of an ancient structure which had in its time been used as a State House, and for many other purposes.

The county jail is a fine, substantial building, erected a little earlier than the court-house, and the grounds are adorned with fine fruit and ornamental trees. Most of the State institutions are located at Concord.

In 1812, the State Prison was built; but twenty years later it was found necessary to enlarge the original building, and the north wing was added. It contains about one hundred and twenty cells, some of which, it is to be hoped, will still continue to be unoccupied for many years. It contains the cook rooms, which are operated by steam, a hospital, and a hall where religious services are held. There is also a library of one thousand volumes belonging to the institution. The prison yard comprises two acres, and is securely enclosed with a substantial granite wall. The workshops are convenient, and the affairs of the institution have generally been so well managed as to make it self-sustaining.

The Insane Asylum is a fine building, properly arranged in apartments suitable to the sexes, and to the degrees of insanity which afflict the unfortunate patients. The best of medical aid and experience, careful nursing, and attention to comfort are provided as well for incurables as for those whose cases are of a more hopeful character; and it is pleasant to know that when reason is dethroned by some unusual or extraordinary mental strain, the state of intellectual chaos may not be perpetual or hopeless, but is often permanently cured by the treatment received at this institution.

The common schools of Concord bear a favorable reputation; the school-houses are commodious, and some of them costly structures. The facilities for a thorough education in the common and some of the higher branches are excellent. For many years past the whole State of New Hampshire has been waking up to the importance of providing the most liberal facilities for common-school education. Concord, where the representative men of the State annually assemble, and display the talent and learning which are so generally diffused, has, perhaps, been more thoroughly aroused by this circumstance, and has sought to place her common-school system on a high and progressive basis. The schools are graded, as in most large towns, for greater convenience and efficiency in the process of instruction, which ranges from the elementary principles to the higher branches of English education.

The prosperity of Concord has, undoubtedly, been much increased and its growth accelerated by the completion of many railroad lines which centre here, and the population is estimated at sixteen thousand.

The Concord Railroad was open for travel September 1st, 1842; the Merrimack and Connecticut River Railroad was open for travel September 20th, 1849; the Northern Railroad in 1846. The Boston, Concord, and Montreal Railroad was opened May 10th, 1848, although the line through was not completed until five years after. There is also the Contoocook Valley, the Concord and Claremont, and the Concord and Portsmouth. The enterprise which has constructed and equipped so many railroads could not fail to manifest itself as it has in the rapid advancement of Concord, nor could these roads fail each in its way to add steadily to the wealth, business, and prosperity of the city.

The immense travel over these roads, the frequent assembling here of conventions and other large political bodies, as well as other organizations, the assembling of the General Court and many strangers on business connected with the State institutions, necessitates the maintenance of large hotel accommodations, and there are some eight or ten in the city, several of them first-class hotels in every respect.

There are fifteen churches and twenty-five clergymen in Concord; still this large number by no means proves that tares grow rank and plentiful in the moral wheat fields, as some of them are editors, professors, and chaplains of the different State institutions.

There are about thirty lawyers; but this large proportionate number may not be taken to demonstrate that the community is greatly given to litigation, as Concord is the head-quarters of the legal profession in the State, and perhaps the aid of some of them is required to assist in mystifying legislative enactments. There are only some twelve or fifteen physicians, representing, however, all of the various methods of practice; which is a favorable indication of the sobriety, regularity, and superior sanitary condition of the population generally.

Seven or eight newspapers are published in Concord, representing the different religious and political opinions, and most of them are generously supported. The most extensive and famous carriage-manufacturing business in the State has been carried on for many years in this place by several companies. Every kind of vehicle is turned out, and in the highest degree of perfection, combining style, durability, and finish. These carriages find a ready market in all the States and territories of the Union, Mexico, South America, Australia, and many other places. There is also an extensive granite quarry, and "Concord granite" is well known in many parts of the country, as much of it is shipped to distant markets.

In the flourishing village of Fisherville, situated at the confluence of the Contoocook and Merrimack Rivers, is an extensive cotton manufactory with a capital of five hundred thousand dollars. There are also woollen mills in Concord, employing altogether several hundred hands.

Among the many able and distinguished men, native and resident, it may not be invidious to briefly mention a few: The Rev. Timothy Walker settled in Concord, then Pennacook, in 1726, and was pastor of the First Congregational Church, organized in the town, for fifty-two years. He was strong and active in his efforts for the good of the community, successful in his labors, a wise as well as a good man. He lived to see the rejoicing over the surrender of Cornwallis at Yorktown, and the independence of his country acknowledged, when he exclaimed, "It is enough! 'Lord, now lettest thou thy servant depart in peace!'"

Hon. Timothy Walker, his son, was also intellectually a strong man. He graduated at Harvard College in the class of 1756; was a member of the Committee of Safety in 1776; also of the Constitutional Convention in 1784, and held many other offices of honor and trust. He was an officer in the New Hampshire forces during the Revolutionary struggle. After the formation of the State government he was for several years a member of the General Court, and for a long period occupied the position of Chief Justice of the Court of Common Pleas. He died in 1822, at the mature age of eighty-five years.

Benjamin Thompson, long a resident of Concord, married a daughter of the Rev. and a sister of the Hon. Timothy Walker, and is better known as Count Rumford. About 1775 he went to England, where he was intrusted with the position of private secretary to a distinguished nobleman, who, pleased with his intelligence, his superior capacity for business, and his fidelity to the trust reposed in him, procured for him the commission of colonel in the British army, where he served until 1784, when his scientific inquiries, investigations, and attainments attracted the attention of Europe, and he was promoted to the position of lieutenant-general of dragoons in the service of the Duke of Bavaria. He distinguished himself as a soldier and philanthropist, and was honored by the duke with the title of count, afterwards receiving from England the honor of knighthood. He died in 1814. The Countess Rumford, at her death, made very munificent bequests for charitable purposes in Concord.

Franklin Pierce was born in Hillsboro', New Hampshire, November 23d, 1804. He graduated at Bowdoin College in the class of 1824, and, having decided on the profession of law, he entered the office of Judge Levi Woodbury, at Portsmouth, and finished his preparatory studies by two years in the office of Judge Parker and at the law school in Northampton. In 1827, having been admitted to the bar, he commenced the practice of his profession at Hillsboro', in a dingy office which was finished off for the purpose in the end of a shed; and thus, like most men who have acquired distinction and renown in this country, the path had a humble beginning which led him from obscurity, through long years of effort, toil, drudgery, and trials, to reach it.

It seems to be a wise provision of Providence, that what is not worth making an earnest effort to obtain is not worth having, even though it is only daily bread; and the effort required to provide either food or achieve a just fame properly qualifies an individual to appreciate its benefits and realize and discharge its obligations and duties. Thus, when a man enlists with a full determination to serve faithfully and well, and rises from the humblest rank to the highest command, for no other consideration than diligent and efficient services, it would seem a self-evident proposition that he is a gallant soldier and an excellent officer.

His first case was a failure, — no doubt to him a mortifying one; but it neither weakened his ambition nor relaxed his efforts to compel success: on the contrary, it was a spur which goaded him to more extensive preparations and a marshalling of hitherto undeveloped faculties, and it is more than possible that a discomfiture in the outset aroused his latent energies and was a guaranty of ultimate triumph. At the age of twenty-five he was chosen to represent his native town in the Legislature, — his father having been elected governor the same year, — and served four years in that body, the two last being elected speaker by a majority of one hundred votes.

In 1833 he was elected to Congress, being the youngest, one of the most industrious and useful members. It was said of him that he never talked for Buncombe; but his sterling abilities as a watchful and efficient legislator were practically recognized by the House in placing him on some of the most important committees, as well as by his constituents.

He was a disciple of Andrew Jackson, and the warmest personal friendship and esteem existed between them. Here he served four years, and the satisfaction which he gave in the discharge of the duties of a representative may be inferred by the fact that he was then promoted to the Senate, and took his seat in 1837, at the commencement of the administration of Mr. Van Buren, and participated in the conflicts of the intellectual giants of that most brilliant epoch of the United States Senate since the early days of the Republic. Starting in life with a commendable ambition, which was rather to excel in every position he occupied than to gain others still higher, in all respects his career had been one of unexampled success; official station having sought him out and advanced him in every instance at the earliest moment that age would allow him legally to accept it; and now, having reached the goal of his ambition, and having made an honorable record in that great arena of intellectual gladiators, his ambition sated, he determined to resign and retire to the pursuits of private life. Accordingly he resigned the office of senator in 1842, and, returning to Concord, where he had established himself in 1838, resolved to provide some pecuniary means by the practice of his profession to serve his family in an emergency, as his inflexible honesty had left him, at the voluntary termination of a long and valuable public career, a poor man.

On the assumption of the presidential office by Mr. Polk he wrote to Franklin Pierce, urging him to accept a position in the new cabinet. In his communication the president said: —

"In tendering to you the position in my cabinet I have been governed by the high estimate which I place upon your character and eminent qualifications to fill it."

The following is an extract from his reply declining the appointment: —

"Although the early years of my manhood were devoted to public life, it was never really suited to my taste. I longed, as I am sure you must often have done, for the quiet and independence that belong only to the private citizen; and now at forty I feel that desire stronger than ever. Coming so unexpectedly as the offer does, it would be difficult, if not impossible, to arrange the business of an extensive practice between this and the first of November in a manner at all satisfactory to myself, or to those who have committed their interest to my care, and who rely on my services. . . .

"When I resigned my seat in the Senate, in 1842, I did it with the fixed purpose never again to be voluntarily separated from my family for any considerable length of time, except at the call of my country in time of war. . . .

"These are some of the considerations which have influenced my decision. You will, I am sure, appreciate my motives. You will not believe that I have weighed my personal convenience and ease against the public interest, especially as the office is one which, if not sought, would be readily accepted by gentlemen who could bring to your aid attainments and qualifications vastly superior to mine."

A few years after, the contingency unexpectedly occurred to which he had referred in the above letter, by the breaking out of the Mexican War, and the name of Franklin Pierce was the first enrolled as a private in the first volunteer company raised in Concord for the war, and in the preliminary drills he was punctually in the ranks with his musket. He was appointed colonel of the New England Regiment, and the same year (1847) was commissioned brigadier-general, and was ordered with his command to Mexico, where he arrived about the first of July, in the midst of the sickly season, when the vomito was an efficient ally of the enemy. During the campaign which terminated with the capture of the city of the Montezumas, Gen. Pierce was conspicuous among the volunteer officers for prompt and decisive action, regard for the comfort of his men, kindness and personal attention to the sick and disabled, and his temerity and disregard of consequences which might result to himself personally in times of great peril.

In the fierce struggle of contending armies in the valley of Mexico he was constantly under fire, and the many bullet-holes through his clothing attested his hair-breadth escapes. On one occasion, at Contreras, August 19th, he met with a severe accident, his horse falling on the battle-field and crushing him severely, leaving him exposed and defenceless under a galling fire.

When Santa Anna proposed an armistice, with a view to peaceful negotiation, Gen. Pierce was selected by the commander-in-chief as one of the commissioners, and at the triumphant close of the war he returned to Concord, where he at once resumed the practice of his profession, determined henceforth to enjoy that independence which he had known only as a private citizen.

Again, in 1852, a command reached him from a source which he had never disregarded or disobeyed. The people called him to the most exalted and responsible position on the face of the earth ; and, though his opponent was the most renowned military chieftain of the age, and was held in affectionate regard by the American people for his high character and distinguished services, Gen. Pierce was elected with unprecedented unanimity, only four States declaring for his opponent; and thus, before he had reached his forty-eighth year, he was president-elect of this great nation.

During the campaign which resulted in his elevation he encountered obloquy, vituperation, and even calumny, unsurpassed, perhaps, by the fierce and heated canvasses of Jefferson and Jackson, while his administration was assailed, as all administrations before and since have been, with or without cause. Still it may be considered as proof, as far as it goes, that this persistent and malignant opposition was groundless, as retrospection and the lapse of time have softened political asperities, and a land burdened with grief, taxation, debt, and sectional animosity would rejoice at the return of those palmy days of the Republic.

Gen. Pierce being remarkably urbane and courteous by nature, as well as from a sense of duty, to all men, friend and opponent alike, without regard to station or position, very properly, when it is his legitimate and unquestioned right, exacts it for himself; and, recently, when a person in high official station offered him a gross and premeditated indignity, it afforded even his opponents pleasure to see him — only a private citizen — bring the official offender promptly and humbly to his knees.

Ex-President Pierce has been a resident of Concord for the past thirty years, and has been uniformly an active promoter of all its interests. Generous to a fault, he dispenses his moderate means in the most effective way to alleviate suffering and want, and sad hearts, filled with gloom and dark despair, are illuminated by his sympathy and gladdened by his material bounty.

Hon. Isaac Hill was a resident of Concord for more than forty years. He commenced life as a journeyman printer, but rose to be an editor, and wielded a powerful political influence in New Hampshire for many years. Energetic, enterprising, and benevolent, most decidedly democratic in all his notions, and possessed of good intellectual abilities of the positive kind, he vaulted into popular favor and was honored with many positions of trust and profit, among which may be mentioned State and United States Senator and Governor of New Hampshire. He died in 1850. Ex-Governor Kent, of Maine, a gentleman of superior abilities, was also a native of Concord.

That Concord is not a transient place is shown by the fact that the descendants of the first settlers still constitute a considerable portion of its inhabitants, such as the Bradleys, Stickneys, Eastmans, Walkers, Rolfes, and others, many of them still occupying the same estates, and recognized as among the most prominent of the citizens of Concord.

In 1853 the town of Concord adopted a city charter, and Gen. Joseph Low was the first mayor.

Several bridges span the Merrimack at Concord, — the first at Fisherville, on the highway; the Boston, Concord, and Montreal Railroad bridge; the upper free bridge, anciently known as the "Federal" bridge; the lower free bridge, on the highway to Pembroke; and the Concord and Portsmouth Railroad bridge.

The Merrimack River is a broad, placid stream from Garvin's Falls, four miles below Concord, to the foot of Webster's Falls, at the forks in Franklin, with the exception of Sewall's Falls towards the north part of Concord. A dam was constructed at Garvin's Falls a few years since by the Amoskeag Manufacturing Company, more particularly to avoid losing their charter, than with a view to commencing manufacturing operations at present.

At the foot of Sewall's Falls is Sewall's Island, formerly known as Pennacook Island. It was once a famous resort for the Indians, and is said to have been the principal residence of the Sagamon or Bashaba.

Many years ago, when manufacturing upon the Merrimack had begun in earnest, surveys were made and measures taken to apply the water-power at Sewall's Falls to such use. The track of the canal was marked out, and some excavations made, which can still be traced; but the fall being inferior to many others on the river, and there being difficulties in the way of constructing a dam, the project was ultimately abandoned.

Sometimes when the snows dissolve upon the mountains, and the great spring floods come down, the river overflows its banks and deluges the broad intervals, presenting the appearance of a vast lake, and often doing great damage, but fertilizing the lands by irrigation and the sediment which is deposited. The Soucook River, which rises in Gilmanton, falls into the Merrimack at Concord on its east bank. Much of the land bordering this stream is a light pine plains soil; hence, no doubt, its Indian name.

Bow was granted, in 1727, to Jonathan Wiggin and others. Long before the introduction of railroads into the State a canal was constructed around Garvin's Falls, called the Bow Canal, to facilitate the navigation of the Merrimack. This canal is twenty-five feet in perpendicular height, and about one hundred and twenty rods in length, cut for the most part through solid granite, and was well calculated for durability. Turkey River, the principal stream in this town, empties into the Merrimack at Turkey Falls, a mile above Garvin's. These falls have likewise been improved, making the navigation less difficult.

Samuel Welch died in Bow, April 5, 1823, at the remarkable age of one hundred and thirteen years. He was born in Kingston, September 1st, 1710, but had been a resident of Bow for fifty years previous to his decease. His life had been a quiet one; he was temperate and industrious, and cultivated his farm until he was a hundred years old.

Hooksett was originally called Harrytown, and was embraced in several of the early grants. It was finally taken from the territory of each of the towns of Chester, Goffstown, and Dunbarton, and incorporated under its present name in 1822. Brick-making is extensively carried on, the clay being peculiarly adapted to this purpose, and several millions of brick are made annually. Among those early engaged in this business was Hon. R. H. Ayer, then a resident of Hooksett, who was extensively and successfully engaged in making brick for some years. Mr. Ayer was a gentleman of the old school, a distinguished son of New Hampshire, and a warm personal friend of Daniel Webster. He died about twelve years since at Manchester, where he had resided many years, possessed of an ample fortune, accumulated by a long life of enterprise and industry.

The "Pinnacle" is a high rock rising abruptly some two hundred feet or more from its base, at which there is a beautiful pond, clear, and of great depth; the indications are strong, and to some conclusive, that by some violent convulsion of nature this rock was upheaved from the place now occupied by the pond. From the Pinnacle is obtained some of the finest specimens of crystallized granite quartz.

The Isles of Hooksett Falls, in the Merrimack, at Hooksett Village, are romantic and abrupt, — a high rock standing in the centre, the fall being sixteen feet in a few rods. There is an extensive cotton manufactory owned and operated by the Amoskeag Company of Manchester.

To facilitate the navigation of the Merrimack, a canal was constructed some forty years since around these falls ; it was about eighty rods long, and cost over six thousand dollars. Two bridges span the river at Hooksett Village, — one on the highway, the other on the Concord Railroad.

Some years ago silver and lead ore were discovered, and as there were indications of extensive mines, a company was formed to operate them.

CHAPTER VIII.

Amoskeag Falls. — Indians. — Fisheries. — Manchester. — History of its Manufactures. — The Starks. — City Institutions. — Samuel Blodget.

ON leaving Hooksett Falls, the back-set of the Manchester dam is at once observed, and for the distance of eight miles before arriving at the Amoskeag Falls, the river is broad, placid, and beautiful; not a ripple ruffles its surface. It is bright, clear, and deep, and about three hundred yards in breadth, its banks, either high and steep and fringed with foliage, or broad, rich intervals, while the gentle swells rolling back from the river, high and still higher, like emerald waves, are dotted with neat farm-houses, orchards, and cultivated fields; the great hills and the distant sombre forests all combine to give this section of the Merrimack River and its surrounding territory an air of pleasantness and thrift. A few years since a fine steamboat plied between Amoskeag and Hooksett Falls for the accommodation of excursion and pleasure parties; but the admonitions of economy have put out its fires. More recently, however, the Hon. E. A. Straw has run a beautiful steam yacht on his own account.

Two hundred years! and how changed the scene! From a desert wild over which the rude barbarians held despotic sway, where nature enveloped in a primeval panoply of shade could not, nor was she invited to, display the diversity and perfection of her features, or the generosity of her bounty and good will to man, by a prospect like this throughout the red man's vast domain. Two hundred years have rolled the vicissitudes of time over New Hampshire, and marked their decades as strongly and visibly in her physical change and improvement, as in the march of intellect and the progress of civilization. At that time interminable woods, clothed in luxuriant foliage, shut out the germinating rays of a genial summer sun, and condemned the soil to worthlessness and waste, while the vast water-power of the Merrimack swept down over the magnificent falls of the

Namoskeag without molestation, and they who owned this land, now so adorned and teeming with prosperity and wealth, lived upon it in wigwams of squalid discomfort, and in vagrant idleness, arrayed, if at all, in furs wrought in fantastic form, or in the picturesque costume of a single ornamented blanket, paint and plume of fancy-colored feathers, depending on the ashen bow, the rude spear, and the ahquedauken for their sustenance, with no thought, or even an idea, of progress or elevation above the wretchedness of this brutish and vagabond life.

The indications of the great change which two centuries have made are forcibly presented to the external senses in the substitution of clear and cultivated fields, in binding and detaining the mighty waters of the Merrimack, and applying its great power to the benefit and comfort of mankind; in the beautiful city where industry and thrift are companions; where pavements resound with the tumultuous rush and rattle of busy strife in place of the stealthy trail of the lurking savage; where monster mills send forth the pleasant hum of industry to take the place of the war-whoop; and whose neat, convenient and comfortable dwellings have superseded the comfortless hovel of the barbarian; while the iron horse, with almost interminable trains, rushes along the river bank, — as much of an improvement on the frail bark canoe as the electric telegraph now strung along the river is on his line of fleetest runners.

Namoskeag was a famous resort for the Indians. Indeed, it was the home of a tribe whose confederates assembled annually around this great fishing-place, and were welcomed with feasting, revelry, and savage sports. Here also came old " Papisseconewa " (the pappoose or child of the bear), the principal sagamon, the inferior chiefs, prophets, medicine-men, and other dignitaries of the confederate tribes, who watched the progress and success of the fishermen, discussed and arranged the affairs of the confederacy, and indulged in or witnessed the grotesque entertainments customary on such occasions with stolid Indian gravity; and though perhaps not so demonstrative, evidently enjoyed the entertainment provided, and the hospitalities of their hosts, with as much freedom and gusto as the brave, and squaw, or the enthusiastic and irrepressible youth.

"The Namaoskeags resided at the falls in the Merrimack known at present by the name of Amoskeag in Manchester. This word, written variously, Namaske, Namaoskeag, Naumkeag, and Nainkeak, means fishing-place, from namaos (a fish) and auke (a place)."

Of the three great principal-fishing places on the Merrimack, Pawtucket, Namoskeag, and Winnipesaukee, Namoskeag was considered the best, probably because of the greater facility of taking fish afforded by the conformation of the falls. The water rushes furiously through several narrow gorges or channels between the rocks, the least rapid being on the west side, and falls into a large eddy, and to this they most likely crowded, as the rush of water from above, and the fish from below, prevented the ascent of large numbers, until after repeated efforts they would fall back into the eddy below, and the Indians, having a weir or kind of seine around the eddy (except a narrow opening on the side next to the falls for their admission), could take them at will.

In the fishing-season the Indians were engaged in taking the fish, the squaws preparing them for the table, for preservation, and for barter, while the night was passed in dancing and feasting. At these fishing-seasons lovers' vows were plighted, marriages were consummated, speeches made, and treaties formed. Particular periods and important events were marked and celebrated among them with great ceremony. Among the established institutions, or customs, may be mentioned the Recruiting or Fire Brand Dance, when the declaration of war had been announced by the chief.

> " By the red sun's parting glance
> They gathered for the warrior's dance;
> First in a circle wide they stand,
> Each with an arrow in his hand;
> Then crouching, and with bended bow,
> They step to measure light and slow;
> Now quicker with a savage flurry,
> They circle round and hurry, hurry;
> Now the ring breaks, and leaping, yelling,
> In one discordant chorus swelling;
> Then tomahawks are brandished high;
> Their shouts re-echo from the sky;
> Their blood-stained nostrils opened wide;
> Their foaming lips all dark and gory,
> Make up the red man's scene of glory." — STARK.

"Brushwood, pitch-knots, clubs, and sticks were gathered in an immense pile near the wigwam of the sagamon. He and his principal chiefs formed a ring around this pile of brush, sitting cross-legged upon the ground; next to these the warriors formed a second ring, and back of these the old men, women, and children were mixed without order or rank. The pile being fired in due time, the principal chief stepped into the ring, and dancing around flourished his tomahawk and knife, naming his exploits, and the people with whom he was at enmity. At the mention of every enemy he would strike the fire with his hatchet, seize a brand, flourish it about in numberless vibrations with his hands, and, contorting his body into every conceivable shape, he would bury his hatchet deep in the ground and leave the ring. Others would follow, and in the same manner dance about the fire and fight it, closing with burying their hatchets in the ground, till the whole of the warriors inclined to follow the war-path had joined in the dance. Every man who joined in the dance was considered as enlisted for the war." This ceremony was always performed in the night; but when the exigencies of the case required immediate action, a ceremony of the same signification was performed in the daytime.

"Around a sapling in the grove, or one standing near the wigwam of the sagamon, after relating their adventures as they danced about the ring, each warrior closed his dance by striking his tomahawk into the sapling; and every one who struck the sapling was claimed as a volunteer upon the war-path. The chief then appointed his rendezvous, and the warriors repaired to their wigwams to make their slight preparations for their departure."

The Scalp Dance was performed by the braves on their return home from the war-path when success had crowned their arms.

"Each Indian hung to his girdle the scalp locks he had taken in other wars, if he had taken any, while the fresh scalps were held by the hair between his teeth. The Indians, thus garnished with these horrid trophies, took a stooping posture, so that the scalps suspended from their teeth might not touch their bodies, and in such positions commenced the most hideous cries and furious stamping, jumping, and dancing about like mad men, ever and anon taking the scalps from their teeth to recite the incidents connected with the killing of the enemy, and then replacing them, to continue the frantic dance with redoubled fury.

"These dances were truly horrible, and led Nathaniel Segar, who witnessed one in 1781, on the sources of the Amariscoggin, as performed by Tom Hegon and his party, to this quaint and laconic description. 'Their actions are inconceivable. It would seem that Bedlam had brooken loose, and that h—l was in an uproar.'

Sometimes a drum was improvised by taking a section of a hollow log, and stretching a deer-skin over the end; this beaten lustily with a stick, in connection with the chanting of singers, served for a quadrille band.

The city of Manchester has been granted and regranted, in whole or in part, many times, and under many different names. It was first incorporated September 3d, 1751, under the name of Derryfield; but its settlement was commenced as early as 1720 around the mouth of Cohas Brook and Goffe's Falls, and in the neighborhood of Amoskeag Falls. Some of these early settlers came from Massachusetts, while others came from the Scotch-Irish settlements at Nutfield or Londonderry. This last-mentioned colony was composed of resolute and sturdy men, most of whom had descended from those stubborn Scotch Protestants who had fled to the north of Ireland to escape persecution. In 1718, hearing of a land where religious toleration was the rule, one hundred and twenty families, severing every tie that bound them to the Old World, embarked on the trackless ocean in search of it, and landed safely in New England. In the following year, having procured a title to this tract in the "Chestnut country" from Mr. John Wheelwright, grandson of Rev. John Wheelwright, to whom all this territory had formerly been deeded by Passaconaway and his associates, sixteen families, under the lead of Parson McGregor, settled in Nutfield, giving it the name of the county from whence they had emigrated.

"They were Presbyterians. They lived in that age of enthusiasm when the adherents of old and new creeds gloried in the name of martyrs, and dissenters demanded (what they were seldom willing to grant) unlimited freedom of religious opinion. These emigrants were proud to enjoy and gloried in vindicating the Presbyterian faith. They were descended from men by whom that doctrine had been maintained with a spirit of independence unequalled in any State in Europe, and hardly surpassed by the firmness and valor by which their more remote ancestors, unawed by the terror of the Roman name, defended their moors and marshes against the conquering arms of Agricola."

Many of those who composed this colony were men of mark and note in Ireland, and became eminent historic characters in this country. Among those were John McNiel, who settled near the falls, after reaching Londonderry. He served in expeditions against the Indians, and, being a man of great frame, six feet and a half in his moccasins, well proportioned, athletic, and powerful, he was feared and avoided by the red man and by the white who had incurred his displeasure. Gen. John McNiel, a distinguished officer in the war of 1812, was his great-grandson, and his counterpart in physique.

Gen. McNiel retired from the service in 1830, was appointed surveyor of the port of Boston, and held the office until his death, which occurred at Washington, D. C., in 1850.

Archibald Stark also settled near the falls, and his second son, John, was the victorious brigadier at Bennington, meeting and defeating a superior force of Burgoyne's army. Archibald Stark was an educated man, and a prominent actor in the stirring events of that period. He built a fort for the protection of the colony, near what is now known as Nutt's Pond, near the residence of Rodnia Nutt, Esq., father of "Commodore Nutt," a mile from the City Hall, which was called Fort Stark. He died in 1758.

John Goffe settled near the falls which bear his name. He was a colonel in the service of the province, and a man of considerable note. The house in which he resided was repaired the present year, when it was found that the roof had been originally shingled with birch bark.

The fisheries at Amoskeag were an important interest to the settlers, as they had previously been to the Indians, and for many years hundreds of men came together at the falls and drew largely from them for the subsistence of their families, and though so many, there were fish enough for all comers. One man had equal rights with another; the rule which secured the rights of each being tacitly understood and generally respected, any infringement being settled on the spot by what was termed "Scotch argument."

"The view of the fishermen while on duty at the falls was a scene of no small interest, — a hundred men in their torn and ragged costumes; some half hid in the surrounding gloom, others conspicuous on lofty rocks, which reflected the gleam of their watch-fires, moving in every possible direction, and with every variety of motion, throwing the 'scoop-net,' handling the 'squirming eels,' covered with blood and slime, inseparable from their occupation; some fighting, and all shouting at the top of their voices, — formed a scene worthy of Pandemonium itself. I suppose we have no idea of the immense numbers of fish with which this river once abounded. My father has seen the shad so thick as to crowd each other in their passage up the falls to gain the smooth water above, so that you could not put in your hand without touching some of them, and yet there were more alewives than shad, and more eels than both. It is no wonder eels were called 'Derryfield beef,' for I have heard those say who would be good judges in the matter, that eels enough were salted down in a single year to be equal to three hundred head of cattle. There was one great advantage about the 'lamprey eel.' It had no bones except in the head, and, as that was never eaten, it made safe food for the children." *

This was after the fishing was confined to the immediate vicinity of the Falls, and long after the ancient " Namoskeag " had been reduced from its original proportions to the falls and vicinity. It appears that the Namoskeag of the Indians extended from, and included, Goffe's Falls to Concord, thus being made up of Goffe's, Short's, and Griffin's, Hooksett, Merrill's, Amoskeag, Garvin's, and Turkey Falls; and their fame early extended even beyond the limits of our own country, and a curious communication from Cotton Mather, referring to some of the natural curiosities still to be seen at the Falls, was published in the " Philosophical Transactions," a scientific publication of London, which reads thus: —

" At a place called Ammuskeag, a little above the hideous falls of Merrimack River, there is a huge rock in the midst of the stream, on the top of which are a great number of pits, made exactly round like barrels, or hogsheads of different capacities, some so large as to hold several tons. The natives knew nothing of the time they were made; but the neighboring Indians have been wont to hide their provisions in them in the wars with the Maquas. God had cut

* Extract from remarks of William Stark, Esq., at Manchester Centennial Celebration, October 22, 1851.

them out for that purpose for them. They seem plainly to be artificial."

This is an illustration of the explosion of mere theories, showing that the belief of saint and sinner alike, unsustained by evidence, will certainly dissolve before the bright rays of the sun of knowledge and of truth. Some idea of the importance of this place, and the immense amount of fish taken, may be had from the fact that a cart full could be had for a pint of "Old Medford;" shad, often for hauling away, to make room for dressing salmon and eels. Some slight idea may be had by the following item from the "New Hampshire Gazette," May 23, 1760:—

"One day last week was drawn by a net, at one draught, two thousand five hundred odd shad fish out of the River Merrimack. Thought remarkable by some people."

As fish became less plentiful, certain points on and about the Falls were the most desirable positions. These places were held for the time being by the right of possession, and the claim was not disputed.

Indian relics have been collected in the vicinity in endless numbers and great variety, several large collections being in the possession of individuals, and there is also a small assortment at the State Reform School. Mr. Samuel Kidder has a very extensive assortment; and the collection of Ezra Huntington, Esq., a member of the present and last Board of Aldermen of Manchester, though not so extensive as some, is remarkably excellent in the perfection of the specimens.

There are four bridges across the Merrimack at Manchester: one directly across the Falls; another at the foot of Granite Street; the New Hampshire Central Railroad Bridge, near the mouth of the Piscataquog River; and the Concord Railroad Bridge, at Goffe's Falls.

October 29th, 1727, a severe shock of an earthquake was felt along the Merrimack Valley, probably the most severe of anything that was ever experienced in this country. Chimneys were thrown down, furniture and dishes were scattered about the rooms promiscuously, and the people filled with consternation, many supposing that the end of all sublunary things had surely come, and "those now prayed who never prayed before, and those who had prayed now prayed the more." Tradition has it that the alarm even spread among the brutes; and cattle, terrified at the trembling and oscillation of the earth, ran bellowing about the fields.

About half a mile above Amoskeag Falls the Merrimack River receives a stream known as Black Brook, from the peculiar inky appearance of its waters. Its source is in Dunbarton, and it falls into the Merrimack on its west bank, having a good water-power, and several mills and shops a little distance from the river.

"The Merrimack River passes over Amoskeag Falls, which are the highest on the whole course of the river. These Falls, being in all about fifty-four feet perpendicular height, are justly regarded as a very striking natural curiosity. The river here spreads out to three times its usual width, and is divided into several channels by rocks and small islands. The accumulated waters of the numerous streams, which drain a large portion of the State, rush through the various rough channels into which the river is divided with great velocity, and with a noise that may be heard for many miles.

"Near the upper and greatest falls several circular holes, some of them more than eight feet in diameter, have been worn to a considerable depth perpendicularly into solid rock. It is supposed to have been done by small stones put in motion by the force of the current. In these holes the Indians who formerly inhabited the country around the falls concealed their provisions in time of danger." *

The largest of the wells referred to have been found to measure sixteen feet perpendicular depth. Just below the falls there is an island, on which are several houses and a bleachery, where the cotton goods made in Manchester are bleached.

The first manufacturing business known to have been done at these falls was by the Amoskeag Cotton and Wool Factory, which commenced operations some time about 1809, and appears to have been owned and operated by Benjamin Pritchard; but, soon after, it seems, for some reason, Pritchard made it a joint stock company, an extract from the records reading as follows: —

"At a legal meeting of the Directors of the Amoskeag Cotton and Wool Factory, being duly notified and holden at the house of Robert McGregor, Esq., in Goffstown, March 9th, 1810, present James

* Geography of New Hampshire.

Parker, Samuel P. Kidder, John Stark, Jr., David McQueston, and Benjamin Pritchard."

There were several other directors, whose names appear at subsequent meetings; the name of Jotham Gillis being signed as clerk. In 1813, Frederick G. Stark, Esq. (the late Judge Stark), was appointed agent of this company, with a salary of one hundred and eighty dollars per year.

The factory was nearly forty feet square and two stories high, and was situated midway between the head and foot of the falls, directly below the west end of Amoskeag Bridge.

The cotton used in this factory was parcelled out among the families of the neighborhood to be ginned at four cents per pound, and the yarn was woven by hand at the various houses where women were the fortunate possessors of a loom.

"I have examined the accounts, kept in the beautiful round hand of Judge Stark, for the month of October, 1813, for fifteen days in succession. During that month, there were manufactured at Amoskeag, three hundred and fifty-eight skeins per day of cotton yarn. This was about the average amount: this three hundred and fifty-eight skeins at factory price was worth twenty-nine dollars and twenty-two cents." *

In 1826 the mill was enlarged, and the foundation was put in for another, which was subsequently built, on the island. These mills were both destroyed by fire, — that on the island in 1840, — and the old Amoskeag mill was burned in 1848, and has not been rebuilt.

Operations in the manufacture of cotton at Amoskeag Falls having been commenced in 1809, the business continued, though not remunerative, until about 1816, when work was stopped, probably for want of sufficient encouragement. In 1822, Olney Robinson purchased the property and resumed operations; but being inclined to outside speculation rather than to his legitimate business, the property soon fell into the hands of Larned Pitcher and Samuel Slater, for liabilities incurred. In 1825, Willard Sayles and Lyman Tiffany, — of the firm of Sayles, Tiffany & Hitchcock (now Gardner Brewer & Co.), — Dr. Oliver Dean and Ziba Gay, were admitted to a partnership, and, under the new firm, business at once revived. From this period dates the continued prosperity and in-

* Rev. C. W. Wallace, D. D.

crease in the production of textiles at Amoskeag Falls. Their success suggested and encouraged an extension of the business, and through the sagacity and enterprise of Dr. Dean the firm promptly but quietly took measures for securing control of the water privilege and adjacent lands. This was accomplished, the plan of a large manufacturing city was matured, capitalists were induced to subscribe liberally, and in 1831 the Amoskeag Manufacturing Company was incorporated. The property of the firm was owned by the partners in eighths, and was transferred, or merged in the new company at a stipulated valuation, being exchanged for the new stock. The Amoskeag Company also secured the privilege and the property at Hooksett, which had a capital of two hundred thousand dollars; the fall at that place is sixteen feet in perpendicular height, and is rated at a capacity of one hundred thousand spindles. This concern was operated by them until 1865, when the franchise and all the property was sold and transferred by consent of the Legislature to a separate company, and legislative permission was obtained by them to increase the capital stock to one million dollars.

Too much credit cannot be awarded the founders of the Amoskeag Company and of Manchester. Launching, as they did, on this voyage over an almost unexplored and trackless region of enterprise, filled with unseen and treacherous rocks and reefs and counter-currents, and swept by adverse winds, the result has exhibited their courage and sagacity, demonstrated their wisdom and ability, and rewarded their faith and perseverance. Embarking their fortunes, their best efforts, and their hopes in the scheme, they permitted no obstacle to impede their progress, indulged neither fear nor doubt, relaxed no effort, and were crowned with abundant success. Few corporations, if any, stand higher, either in the celebrity of its products or financial soundness, than the Amoskeag Company, which is of itself a proud monument to the wisdom of its founders, and the integrity, liberality, and ability of those who have had the management of its affairs. It was the good fortune of Manchester to be the natural centre of a populous district, whose agricultural, and other interests it has enhanced, and in turn derived great benefit therefrom. By this reciprocity, which has added largely to her individual interests and enterprise, Manchester has a valuable auxiliary to prosperity and growth independent of manufacturing, which has heretofore been her chief reliance. Of the most eminent men who originated, and were mainly instrumental in rearing what is now, perhaps, in many respects, the finest manufacturing city in New England, Judge Potter, in his very excellent history of Manchester, has given the following brief but interesting sketches. The venerable Dr. Dean is still president of the Amoskeag and Manchester corporations.

"Oliver Dean was born in Franklin, County of Norfolk, Mass., Feb. 10, 1783. After getting a good academic education for the times, he commenced the study of medicine with Dr. Mann, of Wrentham, and finished his studies with Dr. Ingalls, of Boston, graduating at the Medical College of Massachusetts in 1809. He commenced the practice of his profession in Medway the same year; where he continued about a year, moving into Boston in 1810. While in Boston, he married Miss Caroline Francoeur, daughter of John Francoeur, Esq., a gentleman of respectability and wealth, who had fled from France during the French Revolution. Dr. Dean tarried in Boston a year and a half, when, in consequence of the stagnation of business incident to the war, he moved back to Medway, where he continued in the profitable practice of his profession until 1817. In that year, in connection with his brother-in-law, Willard Sayles, Esq., he commenced the business of manufacturing in Medway. He continued in this business with success, until the fall of 1825, when he, in connection with Lyman Tiffany and Willard Sayles, of Boston, and Larned Pitcher and Samuel Slater, of Pawtucket, R. I., purchased the property of the Amoskeag Manufacturing Company at Amoskeag Village, in Goffstown. This property had been purchased, in 1822, by Mr. Olney Robinson, of Pawtucket, who had operated the mills, but to no advantage; and they had mainly passed into the hands of Messrs. Pitcher and Slater, to secure an indebtedness to them for money and machinery. Robinson had made purchases of the large farms known as the 'McGregor farm,' and the 'Blodget farm,' and other real estate, and thus, in connection with his other operations, had become very much embarrassed.

"When the purchase was made of him he expected to have been continued as agent, but it was soon found that he was entirely inadequate to the business, and Dr. Dean was induced, though reluctantly, to take the agency of the company. Accordingly, in the spring of 1826, he moved to Amoskeag and entered upon his duties as agent. This may be considered as the successful starting-point of manufacturing at this place. Possessed of a competent knowledge as a manufacturer, and a man of enterprise and energy, everything about the premises soon began to assume a new aspect. The Bell Mill had been built for a machine-shop, to be under the management of Mr. Ira Gay, of Nashua ; this was immediately enlarged and fitted up with machinery for the manufacture of tickings, in which Dr. Dean had excelled in Medway and Walpole. Soon the works were extended, and the island mill was built, and fitted up for the manufacture of tickings. The tickings manufactured here soon acquired a reputation unequalled, which they retained under the name of 'A. C. A. Tickings.' But Dr. Dean's time was not confined exclusively to manufacturing, — every other interest connected with the prosperity of the town and village came in for his attention. He was a pattern farmer, and the large farm below the falls, and which had become impoverished and overgrown with brambles and bushes, from bad husbandry, soon began to present a new face under his intelligent care, and, from being one of the poorest, came to be one of the best, in the town. The success attending manufacturing under his charge soon attracted the attention of other capitalists, and the project was started of occupying the entire water-power at this place for manufacturing purposes. It is needless to remark that Dr. Dean was the soul of this enterprise. In June, 1831, the Amoskeag Manufacturing Company was chartered, Dr. Dean being the first grantee ; and at the organization of the company under this charter, in July of the same year, he was chosen president of the corporation. He was continued agent and treasurer of the company until 1834, when he declined the agency, having determined to live a more quiet life, and retired upon a beautiful farm in Framingham, Mass. He was still treasurer of the corporation, but declined this office in 1836, and was succeeded by John A. Lowell, Esq., of Boston. He still continued on the board of direction, and, in 1853, upon the death of Joseph Tilden, Esq., he was again chosen president of the corporation, which office he retains at the present time. In 1847, he was chosen president of the Manchester Mills Corporation, and through its various changes has remained a director of the same, and its president. In 1843, Dr. Dean moved from Framingham to Boston, where he now resides. Dr. Dean may be emphatically placed down as one of the Fathers of the City of Manchester, as few men have done more for its existence and its prosperity. Active, intelligent, and communicative; dignified yet courteous; careful of his own interest, yet interested in the welfare of others; in a word, acting upon the principle of 'live and let live,' Dr. Oliver Dean's name stands prominent among the founders of our city, and is ever retained in grateful remembrance.

"The gentlemen now having control of the company had a proper estimate of the capacity of the hydraulic power at the falls, and the ability to avail themselves of its advantages; they therefore very quietly commenced the purchase of the lands in the neighborhood, and were soon the owners of the adjacent lands upon the west side of the Merrimack, that were likely to be needed for extensive manufacturing operations, or that would be enhanced in value, by the building up of a manufacturing city. Gradually, some of the largest capitalists of Boston and its vicinity became identified with the enterprise, and it was determined to commence manufacturing upon an extended scale. Accordingly, in 1831, the charter for 'The Amoskeag Manufacturing Company' was obtained, with a capital of one million six hundred thousand dollars.

"The first meeting of the grantees was holden July 13, 1831, when Oliver Dean was chosen president, and Ira Gay, clerk. The act of incorporation was accepted; the stock of the old company became merged in the new one, at an appraised value, and the remainder of the stock was taken up in a short time. At the annual meeting, July 12, 1832, Dr. Dean was chosen agent of the company. They soon determined to enlarge their operations, and to take means to have their water-power occupied. The plan of this company was to furnish other companies, disposed to locate here, with sites for their mills, and run them upon their own account; and at the same time to put their lands into market in lots for houses, shops, and stores, and thus build up a manufacturing town, at the same time that they greatly enhanced the value of their own property. A reconnoissance by competent engineers developed the fact that the east bank of the Merrimack was the most feasible for their operations, both as a track for their canals, and a site for their mills; and, in 1835, they succeeded in securing the most of the lands upon the east side of the Merrimack that by any possible contingency might be necessary for them.

"Willard Sayles was born in the town of Franklin, County of Norfolk, Mass., in April, 1792. While yet a child his father removed to Wrentham. In 1821, Mr. Sayles moved to Boston, and entered upon commercial pursuits in company with Lyman Tiffany, Esq. They soon became extensively engaged in manufacturing. His first investments in manufacturing were in Medway and Walpole, Mass. At last, as his business extended, he became connected with other establishments, and, among them, with the mills at Amoskeag and at Hooksett. In the latter, he became the largest owner; having, at the time of its being merged with the Amoskeag Manufacturing Company in 1835, twenty-four out of seventy-two shares in that corporation. From this time until the time of his death, Mr. Sayles became largely identified with the operations of the companies in this place. He was a director in the Amoskeag Manufacturing Company, and one of the committee to purchase the lands belonging to that company in this neighborhood. Shrewd in management, and of great tact in driving a bargain, it is probable that few persons could have done the business to better advantage for the Amoskeag Manufacturing Company than he. Many anecdotes are told of his negotiating for the lands, that show his shrewdness in bargaining, if they do not add to his reputation as a man. His operations as manufacturer added to his business as a merchant, and few houses in Boston did a more extensive business in domestics than that of Sayles & Merriam, or to greater advantage. Mr. Sayles died in 1847, at the age of fifty-five, leaving a large property, gained by a life of energy and enterprise.

"William Amory was born April 15, 1804, in Boston. His father, Thomas C. Amory, Esq., an eminent merchant of that city, died in 1812, much lamented. William, his son, entered Harvard University, in 1819, at the early age of fifteen years, and left at the end of three years to finish his education in Europe. He was a student at the University at Gottingen, in Germany, for a year and a half, and at the University of Berlin, for nine months, pursuing the study of the civil law and of general literature. He then spent some two years and a half in travel, returning to Boston May

20th, 1828, after an absence of five years. In 1831, he entered the bar of Suffolk County, without, however, any intention of practising the profession of the law. The same year he was chosen treasurer of the Jackson Manufacturing Company at Nashua, and entered upon the business of a manufacturer. Young and inexperienced in the business, yet he brought to the performance of his duties a mind matured by his study, and a knowledge of men and things, together with an energy and enterprise not often found in young men nurtured in ease and affluence. With such qualifications he was destined to succeed, and the Jackson Mills, for the eleven years they were under the control of Mr. Amory, were eminently successful, and did not fail to make certain and large dividends. In January, 1833, he married Miss Anna P. Sears, daughter of David Sears, Esq., an eminent merchant of Boston. In 1837, when the Amoskeag Manufacturing Company had decided upon building up the manufacturing city of Manchester, they looked about for an efficient man to take the responsible office of treasurer, and of course general manager of the affairs of the company. Mr. Amory's success in the management of the affairs of the Jackson Company very naturally directed their attention to him, and, fortunately for them and our city, he was elected to that office, and accepted it. He entered upon his duties at once, and for nineteen years has been the controlling and directing spirit that has fashioned the destinies of our city. Of just and enlarged views, he has suffered no niggardly policy to mingle itself with the management of the affairs of the company. And the directors have had the good judgment to leave his action unrestricted. And to-day our citizens who take pride in our public library, in our beautiful cemetery, in our spacious streets, in our numerous and splendid public squares, so justly considered as ornaments and sources of health to our city, may attribute them in a great measure to the enlightened policy of William Amory, Esq.; while the corporation, whose business affairs he controls, cannot but appreciate a policy that is fast building up a manufacturing city, unsurpassed in beauty, at the same time that it is creating a stock that in dividends and surplus is the most desirable in the market. Mr. Amory is a stockholder and director in most of the other corporations of the city, and has had much to do with their success. In fact, few men of his age can look back upon such a record of success in manufacturing as the treasurer of the Amoskeag

Manufacturing Company. At the same time that he has been so successful in his manufacturing operations, his urbane manner and quiet and unostentatious courtesy have not failed to secure to him the respect and regard of a host of friends in the cities where his arduous duties have been performed.

"Col. Robert Read was born at Amherst, in October, 1786. His grandfather, Robert Read, moved to Amherst from Litchfield at an early date. His father, William Read, was a well-known and influential citizen of Amherst. Robert was apprenticed as clerk, with Messrs. Haller and Read, merchants of Chelmsford. He afterwards went into trade with his father, at Amherst, under the well-known firm of William Read & Son. Subsequently, he formed a connection in business with Isaac Spaulding, Esq., which continued until a dissolution precedent to the commencement of manufacturing in Nashua, where Mr. Spaulding removed and went into business. Few firms were more enterprising, or better known in our State, than that of Read and Spaulding, of Amherst, and very few were more successful. Mr. Read, active and enterprising, soon became a leading man in the town. He was elected town-clerk in 1815, and was re-elected to the same office the twelve following years. He was also representative of the town for three years; namely, in 1826, 1827, 1828. Prior to this time he had been in command of the West Company of Infantry, in Amherst, one of the most efficient companies in the State, he having served in all the intermediate grades from private to commander. In 1828, Capt. Read was appointed aide-de-camp, by Governor John Bell, with the rank of colonel. In 1835, Col. Read moved to Nashua. In 1837, he was appointed agent of the Amoskeag Manufacturing Company, and entered immediately upon his duties, succeeding Mr. Hartshorn in that office. He continued to perform the arduous and responsible duties of his office for fourteen years, until January 1st, 1852, when he resigned and removed to Nashua, where he has since resided. During the period of his agency, and under his general supervision, a large portion of the operations of the Amoskeag Manufacturing Company, in the way of buildings and real estate generally, were completed. He performed his duties faithfully to the company by which he was employed, as every one will testify who had business to transact with him. Strictly a business man, he mingled very little with other than business men, and hence was very little identified with interests disconnected with the corporations; yet no measure connected with the progress of our city escaped his attention, and he most heartily coincided in all that liberal course of policy, on the part of the treasurer and directors, that has added so much of beauty and value to our city. Col. Read took very little part in the politics of the day after he came to Manchester; still, in 1851, he was a member of the convention for the revision of the Constitution. His first wife was Miss Rebecca French, daughter of Frederic French, Esq., of Amherst. He married for his second wife, Miss Jane Leland, of Saco, Me. His intercourse with his fellow-citizens was marked with that courtesy of manner that ever commands respect. Since his residence in Nashua, probably from want of his usual active exercise, his general health has become somewhat impaired, yet he still lives to take an interest in all the stirring events of our growing city."

From the earliest ages the founders of cities have been regarded as public benefactors, and have received the distinguished consideration to which they have been justly entitled from their fellow-citizens and posterity. Thus Peter the Great derives a due share of his just historic renown from this circumstance, while the Emperor Constantine has at least perpetuated his memory, and tall minarets plainly denote to all men the precise locality of the city which bears his name. If, then, it is eminently proper to bestow credit on such as have, to gratify, perhaps, a questionable ambition, merely by the force of kingly prerogative established a site, how much greater is the title to distinction and enduring gratitude of those who have laid the foundation, broad and deep, of a city like Manchester, and at the same time, by their sagacity, genius, and skill, created and established an interest which should not only attract its population, but also provide them amply with such various labor and business as would ensure each and all a generous support! The founders of Manchester, who have been previously mentioned, were able men in the broadest sense of the term, as the great manufactories which they devised and put in operation, and the fine city they originated, which, though planted in a desert waste, was nurtured by their genius to its present fair proportions, amply testify. But though the city is an offspring of their genius and wonderful forecast, though the result already reached was plainly foreshadowed, and it required no prophetic vision to discern in the character of these men and their comprehensive plan the certainty of ultimate and complete success; still, to accomplish this purpose, to reach this consummation, the services of efficient and skilful coadjutors were required to superintend and carry into practice the important details of this vast and complicated scheme. Without such co-operation they foresaw inevitable failure, and at once cast about to secure the highest order of talent, the most eminent engineers, artisans, and mechanics, men of intelligence and character as executive officers of the various branches of this great enterprise. The city of Manchester being a recently collected community, in other words, being emphatically a new settlement as well as a new city, exhibits unmistakably the well-known characteristics of such a condition of things, as well as other features remarkable and unusual in a young and heterogeneous community. Depending as she has entirely on her manufacturing interests for prosperity and growth; destitute of any and all of those long-established interests or enterprises which move the slow progress of older places, her people having been brought there, rather than born there, it would be only natural to presume that the city in the aggregate and in detail; in its manufactures and its great edifices for that purpose; in its numerous mills and shops, where every variety of mechanical works are carried on; in the location, style, and expense of churches and school-houses; in the arrangement, number, and size of buildings for business purposes; in the architecture, convenience, and adornment of its dwellings, its broad and regular streets, numerous public squares, and all other matters tending to convenience, taste, skill, means, and enterprise of a people attracted here by their inclinations, interest, or some other cause, and who had each for himself marked out his home and surroundings. Observation and the reflection of thoughtful men will, of course, suggest the utter impossibility of so much method, regularity, order, convenience, comfort, and beauty, as the result of such anomalous collection, diversity and incongruity of taste, and will naturally cast about for the invisible cause of all this wonderful system. The skilful and careful perfection of the original city plan; the wise location of the many public squares which beautify and adorn every section of the city; the wisdom which has divided up and distributed fifty acres of common in proper quantities through the city, instead of locating the whole in one body in some inconvenient corner; the care, attention, and cost which have been devoted to the planting of ornamental shade-trees, adorning and beautifying the streets and squares, — one of the most agreeable and comfortable features of the town, to its people, as it is one of the most attractive to strangers; the proper equality in size of building sites; the remarkable absence of the alternation of stores and dwellings seen in many cities; the respectful and convenient distance maintained between the building front and the street line, — these and many other noticeable features are certainly not an exhibition of uniformity of taste and means, nor are they, either, the result of accident. Industry, study, skill, science, genius, in competent hands and under intelligent direction, has devised and accomplished all this. In the year 1838, Mr. E. A. Straw, then a youth of eighteen, having just left school and engaged as civil engineer in the construction of the Nashua and Lowell Railroad, was sent by Mr. Boyden, consulting engineer of the Amoskeag Company, to supply the place of Mr. T. J. Carter, the regular engineer of that company, who was sick, and whose services were needed at that time. Mr. Straw arrived on the 4th of July of the year named, and immediately entered upon the active duties of the position, expecting to remain but a week or so. His engagement consequently antedates the erection of mills, and, indeed, the commencement of all improvements, beyond the initiatory steps, in the city of Manchester, and he may be said not only to be contemporaneous with, but the agent of, all these improvements. When Mr. Straw came to Manchester there was but one principal thoroughfare, running diagonally from north-west to south-east, from Amoskeag village across the old bridge and the plain to the centre of the town. Along this road there were some half-a-dozen houses, among them the Kidder house near the river; another on the site of the residence of William G. Perry, Esq.; one between these two near the track of the Concord Railroad; and a few others. The dwelling on what was then known as the "Young Farm," now known as the "pest-house," having been built in 1834, appeared, externally, much the same as now. Among the first duties devolving on Mr. Straw in the line of his profession was to lay out, according to the plan, for reclamation and putting into market, the lands of the company necessary to accommodate the anticipated rapid increase of population as fast as it should be required for this purpose. Accordingly it was decided to define the north line of Lowell, the east line of Union, and the south line of Hanover Streets, and that the territory thus bounded should at once be brought into market. It may seem a little singular that this street of elegant public and private buildings was, only thirty years ago, a dense wilderness; that the Lowell Street of to-day — perhaps the finest in the city — required, in its opening to Union, nearly a week, and an immense amount of labor in cutting and removing the forest. The land thus bounded was then cleared, burned over, laid out, and sold at auction, October 24, 1838. Every lot was sold; and it may be interesting to those who have subsequently bought land of the company, and made themselves a home in Manchester, to know some of the prices then obtained. The lot on the north-west corner of Elm and Hanover Streets brought fifteen cents and a half per foot; corresponding corner, Elm and Amherst, twenty-six and a half; Elm and Lowell, eighteen and a half; Chestnut and Amherst, nine cents. The next sale, October 1839, the lot at the south-west corner of Elm and Hanover sold for forty-two and a half cents per foot. The lot at the north-west corner of Elm and Amherst was sold the present year (1868), at four dollars per foot.

In three months thereafter (January, 1839) the first dwelling-house was erected, it being the same now standing at the north-west corner of Chestnut and Concord Streets. Having completed some of the mills, together with boarding-houses for the accommodation of the operatives, while at the same time the building of dwellings and stores in the public streets was rapidly progressing, and the manufacture of cotton on an extensive scale was proving a decided success, the company turned their attention to the production of delaines, a business hitherto unknown in this country. For this purpose they commenced operations in the mill at Hooksett, where the new fabric was made; and so far the experiment was a success; but the imperfect knowledge of the art of printing and the totally inadequate machinery for the purpose effectually hampered their progress. The best method of printing worsted goods practised or known at that time was the old system of block printing, and as this was imperfect and slow, the goods were sold as they came from the loom to an enterprising firm in Taunton, Mass., who it was believed realized a handsome profit by printing them for market on their own account. The partial success of the company, however, in this experiment very naturally encouraged a favorable view of this business, and determined them to commence operations as soon as possible on a large scale, and the Print Works corporation was established. It now became necessary to obtain a more extensive and practical knowledge of the art of printing from Europe, and in canvassing for a suitable person to fill this mission Mr. Straw was selected as a discreet, intelligent, and fit agent, and, in 1844, visited the great manufactories abroad, where, under various pretexts and guises, he found entrance to many of the close establishments which so jealously guarded this then youthful and most important art. Having gathered much useful and serviceable knowledge, and engaged some experienced and skilful workmen, Mr. Straw returned the following year. Meantime the parties interested, being substantially the stockholders of the works already in operation at Manchester, had completed the Print Works, and put No. 1 mill in operation, and the company is still heavily engaged in the production of this kind of goods, of a style, quality, and finish unsurpassed by the imported fabrics of older and more experienced foreign manufacturers. Diligent, faithful, and capable in the line of his duties, Mr. Straw grew in the favor of his employers as well as in the respect of his fellow-citizens, and, in 1852, he was selected for the important position of agent of the Land and Water Power Company. Prior and for some years subsequent to this time the Amoskeag Company was divided into three separate departments,— the Land and Water Power, the Machine Shops, and the Mills, — each under the exclusive management and control of its own agent; but experience, if not wisdom, seems to have demonstrated that the triumvirate system was not, and for obvious reasons could not be, particularly advantageous to the proprietors. and as Mr. Straw had discovered superior executive abilities in his position, it was gradually abandoned, and the onerous and responsible duties of each were in turn assumed by him. Of the able and satisfactory manner in which he has discharged the duties of his position for many years, it is not necessary to speak more particularly; but it may be inferred from the material marks of approbation bestowed on him, and the high esteem entertained for him as a manager and a man, by the company. While assiduously engaged in those duties which have brought prosperity to the city of his adoption and profit to his employers, his fellow-citizens have not been unmindful of his course, nor yet of his merits; but, never in any sense a politician, he has steadily and persistently avoided, when possible, the honors and duties of public life, and is an example — rare in these days — of the office seeking the man. Still, Hon. E. A. Straw has not been wholly able to escape serving his fellow-citizens, he having been a member of the Legislature seven or eight years, and once president of the Senate. Few men have by their own exertions met with equal success, or exhibit a fairer record, and it is gratifying to know that his individual prosperity approximates that of the great corporation he manages, and the city in which he resides. The managers of the Amoskeag Company have been men of liberal minds and comprehensive views, and it has been the good fortune of Mr. Straw to possess their entire confidence; consequently the suggestions he has made to them concerning improvements, generally including the present or prospective welfare of the city, have been heartily endorsed by them without hesitation, and thus his great regard for her is plainly traced on every rood of her territory and every section of Manchester. Mr. Straw, as agent of a company which contributes more than a modicum of the city tax, has exercised a controlling influence, always judicious and healthy, over public improvements, inciting and spurring up, if necessary, any laxity of negligent officials, generally by offers of valuable co-operation; and recently, through the paternal and munificent liberality of the company, he has been the agent for presenting the city with a valuable and eligible site for a public library, and has reason to hope to receive contributions sufficient to defray one half the estimated cost of erecting a suitable building for such an institution. As an evidence of the high estimation in which he is held as a practical manufacturer, it may be mentioned that there is an association composed of the agents and treasurers of all the principal establishments in the New England States, of which Hon. E. A. Straw is president. By reference to a publication bearing date of January 15th, 1868, containing the transactions of this association, it will be seen that it is called the "N. E. Cotton Manufacturers' Association;" President, Hon. E. A. Straw, Manchester, N. H.; Vice-Presidents, A. D. Lockwood, Lewiston, Me., Wm. A. Burke,

Lowell, Mass.; Directors, J. S. Davis, Holyoke, Mass., Charles Nourse, Woonsocket, R. I., Phinehas Adams, Manchester, N. H., William P. Haines, Biddeford, Me., Thomas J. Borden, Fall River, Mass., A. M. Wade, Lawrence, Mass.; Auditor, Benjamin Saunders, Nashua, N. H.; Secretary and Treasurer, Ambrose Eastman, Boston, Mass. This Association numbers upwards of one hundred gentlemen, representatives of all the leading manufactories in New England, its objects being the interchange of views and opinions by the presentation of papers, by interrogatories, colloquial debate, publications, and otherwise, for mutual improvement. In the pamphlet referred to, containing the proceedings of the third annual meeting, holden at Boston in January, 1868, is published a very interesting paper read at that meeting by Mr. Straw, on "The principles which should govern the use of drawing as a process in the manufacture of cotton yarns." It will be noticed that in the selection of the officers of this association the great corporations upon the Merrimack are very liberally represented, and no higher testimonial, it would seem, to the distinguished ability of an individual, as a manufacturer, could be desired than that bestowed on Mr. Straw by this association of the most skilful, experienced, and practical manufacturers; and certainly, if it is proper for any man to entertain a just pride in the appreciation of his merits by competent judges, and feel a gratification which perhaps no other circumstance could produce, that man is Mr. Straw.

In 1794, Samuel Blodget, Esq., who had conceived the idea, commenced operations on the Amoskeag Canal, about a mile in length around the falls, to facilitate the navigation of the river. This canal was a stupendous undertaking for that early period; indeed, it was one of the earliest internal improvements of magnitude in the country, and was completed in 1807, at a cost of about one hundred and twenty five thousand dollars. Its projector died the same year, aged eighty-three.

The remains of Samuel Blodget are interred in the beautiful "Valley Cemetery" at Manchester, the lot being surrounded by a group of fine perennials, affording a grateful shade, and the inscription on his monument is as follows : —

To The Memory of The
HON. SAMUEL BLODGET,
Born at Woburn, Mass.
April 1, 1724,
Died at Manchester
(Then Derryfield),
Sept. 1, 1807.
The pioneer of in-
ternal improvements
in New Hampshire.
The projector
and builder of the
Amoskeag Canal.
Erected by
His great-grandson,
JOSEPH HENRY STICKNEY,
of Baltimore, Md.
1868.

Mr. Stickney — grandfather of the above-named gentleman — was mainly instrumental in procuring the change in the name of the town from Derryfield to Manchester.

The canal served a useful purpose for many years, but went to ruin and decay after railroad facilities were introduced along the Merrimack.

The old bridge across the Merrimack, at the foot of Bridge Street, called from its builder "the McGregor Bridge," was said to have been the first bridge ever built across the Merrimack in New Hampshire. It is related that when McGregor mentioned to General Stark his plan of building a bridge, the general remarked that he should have lived long enough, when a bridge across the Merrimack River should be completed; and a gentleman who had crossed the bridge, in 1810, related that he took the toll-gatherer along as a hostage for his safety, and paid his toll after it was insured, at the opposite end. The bridge, or rather the ruins were washed away by a great freshet in the winter of 1850.

Some distance below the site of the bridge there is a large rock in the river exposed, except in times of high water. This rock is known as "Old McNiel" and received the name from John McNiel, who broke through the ice near it, and with remarkable presence of mind and being a good swimmer, struck out for the shore, and having reached a place where he could touch bottom with his feet, and placing his broad shoulders against the ice, with almost superhuman power broke through, and thus saved himself from death by his coolness, presence of mind, and gigantic strength.

Having decided to commence active operations at Amoskeag Falls, and having bought out the proprietors of the Garvin's Falls Water-Power Company at Concord, the Amoskeag Land and Water Power Company in 1836 commenced operations by constructing a substantial dam at the falls with massive guard locks on the east side of the river. From these locks a canal was dug seventy-five feet wide, and narrowing down to forty-five feet, and ten feet deep. This canal is five thousand feet long, and terminates a few rods north of the passenger station, on the Concord Railroad. This is called the upper canal; the lower canal extends from a point near the Stark Mills to the weir below Granite Street.

In 1845, the upper canal was extended to its present terminus, a distance of five thousand feet from the basin. At the same time the lower canal was extended south to the weirs below Granite Street, and north over the track of the Blodget Canal to the Basin, its whole length being seven thousand two hundred and fifty feet.

As the company had, in purchasing the Amoskeag Canal interest, guaranteed to keep it open around the falls, locks were constructed near the site of the McGregor Bridge, and thus communication was still kept open between the canal and the river. The fall, from the upper to the lower canal is twenty feet, and from the lower canal to the river, thirty-four feet.

The Amoskeag Manufacturing Company was chartered in 1831, with a capital of one million six hundred thousand dollars; the old company, on the west side of the river, being the nucleus of the new company. The capital of this company has been increased to three million dollars. Hon. E. A. Straw is the agent.

The Stark Mills were incorporated in 1838; the capital of this company is one million two hundred and fifty thousand dollars, and Phineas Adams, Esq., is the agent.

The Manchester Mills company was incorporated 1839, with a capital of one million dollars. The stock was afterwards increased to one million five hundred thousand dollars, and its name changed to the "Merrimack Mills." In 1851, the name of the corporation was changed to the "Manchester Printworks," and the following year the capital was increased to one million eight hundred thousand dollars. Waterman Smith, Esq., is the agent.

The Blodget Edge Tool Company, with a capital of one hundred thousand dollars, was incorporated in 1853.

The Manchester Locomotive Works were incorporated 1854. Capital, three hundred thousand dollars.

The Seamless Bag Mill has a capital of one hundred thousand dollars. The Langdon Mills, five hundred thousand dollars.

There are five or six very extensive Hosiery Manufactories, and an immense number of shops and mills for almost every conceivable mechanical purpose.

More than half a century since, the people of this town began to realize the immense hydraulic power capable of being applied to machinery by means of the Amoskeag Falls. Many believed the town would some time rival the great cotton city in Lancashire in manufacturing importance, and effected a change of its name, from Derryfield to Manchester, by legislative enactment, in 1810. In June, 1845, an act was passed by the Legislature, incorporating the city of Manchester, and on the first day of August following, a vote of the town was taken on the question of acceptance of the charter. The vote stood four hundred and eighty-five in the affirmative, and one hundred and thirty-four in the negative.

On the nineteenth of August, 1846, the first election was held for officers under the city charter, which resulted in no choice. A month later, a new trial was had, which was successful, and the new city government was inaugurated September 8th, of the same year. The following-named gentlemen have held the office of Mayor of the city of Manchester, since its organization.

1846, Hiram Brown; 1847-48, Jacob F. James; 1849, Warren L. Lane; 1850-51, Moses Fellows; 1852-54, Frederick Smyth; 1855-56, Theodore T. Abbot; 1857, Jacob F. James; 1858, Alonzo Smith; 1859-60, Edward W. Harrington; 1861-62, David A. Bunton; 1863, Theodore T. Abbot; 1864, Frederick Smyth; 1865, Darwin J. Daniels, who died while in office, and John Hosley was elected to fill the vacancy during the remainder of the unexpired term of Mr. Daniels; 1866, John Hosley; 1867, Joseph B. Clark. The Mayor for the current year is James A. Weston, Esq., and the Board of Aldermen is composed of the following-named gentlemen : Ward one, William G. Perry ; two, Ezra Huntington; three, William P. Newell; four, Horace B. Putnam; five, Daniel Connor; six, Joseph Rowley; seven, Chancey Favor; eight, George Gerry. City Clerk, Joseph E. Bennett, Esq.; Treasurer and Collector, Henry R. Chamberlain ; City Marshal, William B. Patten.

As Manchester is a new place, having sprung up, as it were, in a day, it has neither colleges, or other richly endowed, long-established, and celebrated institutions of learning ; but, since the organization of the city government, it has devoted itself to the interest of education with untiring zeal and energy, and the result is, that Manchester is blessed with facilities for education inferior to those of no other town in the State. The first school-house ever built within the present limits of the city was near Amoskeag Falls in 1785, and nothing could exhibit the spirit of progress in a more favorable light than contrasting its cabin-like, seven-by-nine proportions, its inconvenience, its total destitution of comfort and of finish, with the magnificent edifice recently completed for High School purposes.

Another school-house was built at what is called the Centre, soon after the one at the falls, but this most important interest of any community made little or no progress until Manchester became a city; it is now, however, on the flood-tide of prosperity and success.

Moody Currier is chairman of the Board of Education; Joseph G. Edgerly, Superintendent of Public Instruction, and William Little, Esq., Clerk.

The new High School house was completed in 1867, at an expense of fifty thousand dollars, exclusive of the building site; beside this there are about forty school-houses in this city, many of them fine buildings. The number of pupils the past year was four thousand, employing seventy teachers.

There are eight wards in the city, with one hotel and two churches for (not in) each ward. The first church ever built in Manchester was commenced in 1760, and finished, so far as to be fit for occupation, in thirty years. This was at the Centre; but it was not until 1840 that any clergyman was settled in town, which was over the First Congregational Church at Amoskeag.

No religious organization in Manchester exhibits more positive indications of a thrifty and progressive condition than the Catholic.

The first church of this society was erected in 1849, the expense of it being twenty-five thousand dollars; but the society increased so rapidly that it was found necessary to build another, which is to be completed the present year, and will probably be the largest and most costly church in the State.

There is also a Convent of the order of the Sisters of Charity and Mercy, and the visits of these ministering angels, among the poor, the destitute, and the suffering, are not few or far between. These benevolent and cultivated ladies, by devoting their lives, their energies, and their best efforts to the relief of want, destitution, and suffering, illustrate and exemplify the inestimable worth and beauty of practical Christianity.

Connected with this is an educational institution of a high order, — "Mount St. Mary's Academy," — where all, without distinction of creed, may obtain the higher branches of English education, as well as in the fine and ornamental arts.

In addition to contributing their quota — in the way of taxation — to the support of common schools, the Catholics maintain several large Sabbath and secular schools.

The citizens of Manchester are favored with all the advantages and benefits of a free city library, which contains fifteen thousand volumes. This library was established in 1844 by an association of gentlemen; but the necessity of a more comprehensive institution of this kind being apparent, it was, in 1854, transferred to the city under the following obligations: —

"The said city shall annually appropriate, and pay the trustees of the said city library, a sum not less than one thousand dollars, to be expended in the purchase and binding of books and periodicals, not being newspapers; shall, by suitable appropriations, provide for a room, lights, fuel, and other contingencies of the library, and for the salary of a librarian;" and became the nucleus of the present Public Library,* and under the present arrangement to a much

---

* This institution was fortunate in being surrounded by warm friends in its early days; none more ardent and untiring than the late Judge Samuel D. Bell. Many gentlemen contributed valuable additions to it in volumes, and otherwise; also, the Amoskeag Company manifested a lively interest in its growth and efficiency; but Judge Bell, being an antiquarian and a gentleman of erudition, was indefatigable in promoting its prosperity, and to

greater extent supplies the want of a population like that of Manchester. This is not the first library; indeed, it will be seen that one was established in this town at an early period. In the course of his remarks at the centennial celebration, Albert Jackson, Esq., said: —

"In 1796 a social library society was organized, and in 1799 was, by special act of the Legislature, incorporated."

There are several periodicals, monthly and weekly newspapers, and three dailies printed in Manchester, all well sustained; and thus it would seem that this city has provided the means for moral and intellectual culture, united with enterprise and industry, not surpassed by any similar community, and is believed to be one of the fruitful causes of her numbering, at the age of twenty-one, a population of scarcely less than thirty thousand souls.

Manchester is generously provided with public grounds, there being six separate commons, or squares as they are called, on the east side of Elm Street. These are known by the names of the "Park," "Merrimack," "Hanover," "Concord," "Tremont," and "Reservoir Squares;" the latter containing the companies' reservoir, which is located a little more than two hundred feet above the river. It is four hundred and eighty feet long, two hundred and thirty-four feet wide, and eighteen feet deep, its capacity being eleven million gallons. It is supplied by water forced up from the river, and hydrants are conveniently arranged for using it in case of fire at the mills and boarding-houses. The reservoir is terraced, and furnishes a fine promenade, and the square, like all the others, is tastefully adorned with shade trees. These squares contain six or eight acres each, and were the munificent gift of the Amoskeag Manufacturing Company to the city.

A fine brook comes down from Oak Hill and crosses the city diagonally, passing through several of these squares, which are thereby provided with beautiful ponds.

There are several other smaller squares on the west side of Elm Street, which are yet the property of the company.

Manchester has some ten or twelve cemeteries and burying-grounds. The "Valley Cemetery" containing twenty acres, a gift

---

him, more than to any other man, belongs the credit of providing a library which is now established on a basis that insures its being, in a few years, one of the finest in the State.

of the Amoskeag Company, is unsurpassed for its romantic beauty. The "Pine Grove Cemetery" is well adapted for a silent city of the dead. The Catholic Cemetery is on the west side of the river. Among the oldest inscriptions on any of the tombstones are the following: —

> "Here lyes the body of
> Mrs. Janet Riddel, wife to
> Mr. Samuel Riddel. She Died Septr. 18,
> 1746. Aged 50 years."

Another reads thus: —

> "Here Lyes The Body of Mrs.
> Chresten McNiel. She
> Died September 17th, 1752.
> Aged 66 years."

Here is still another: —

> "Here Lyes The Body of Mr.
> Archibald Stark. He
> Departed This Life June 25th,
> 1758. Aged 61 years."

The City Hall was erected in 1845, at a cost of thirty-five thousand dollars. The State Reform School is also in Manchester; the building is four stories high, convenient and roomy, and the lot on which it is built contains one hundred and ten acres.

Manchester, like most manufacturing cities, is eminently a transient place; but many of the descendants of the first settlers still occupy the old estates, and are counted among the most valuable citizens; but, beyond all question, Gen. John Stark was the most renowned of all her native citizens. He died in May, 1822, at the advanced age of nearly ninety-four years, and was buried in a cemetery provided by himself for that purpose on his own farm, on a commanding bluff just above Amoskeag Falls, whose rushing, rumbling cataract, like the roar of mighty battle, in which, from a stern sense of duty, he engaged and won unfading laurels when living, is his perpetual requiem. A plain granite shaft, bearing the inscription "Maj. Gen. Stark," marks the final resting-place of the patriot and hero.

Wednesday, the 22d day of October, 1851, the citizens of Manchester celebrated the centennial anniversary of the incorporation of Derryfield. An address was delivered by the Rev. C. W. Wallace, an original poem by William Stark, Esq., — a genuine descendant of the "old stock," — and several brief, pertinent addresses were made by native and adopted citizens of Manchester. At the same time Hon. Chandler E. Potter was selected to prepare a history of the town, which duty he has performed in a most able, thorough, and satisfactory manner.

CHAPTER IX.

Cohas River. — Massabesic. — Londonderry. — Scotch-Irish Settlement. — Distinguished Men. — Derry. — Piscataquog River. — Francestown. — Weare. — Goffstown. — Bedford. — Souhegan River and Towns along its Course. — Litchfield. — Reed's Island. — Hudson. — Nashua River and the Towns watered by it. — Dunstable. — The Pequauket War. — Nashua. — Tyngsboro'. — Chelmsford. — Stony Brook. — Dracut. — Beaver River. — John Nesmith.

COHAS RIVER, which empties into the Merrimack on its east bank at Goffe's Falls, is the outlet of Massabesic Lake, and is four miles in length. On this stream there are extensive hosiery, lumber, and other mills, the former located near its confluence with the Merrimack. This stream was called Massabesic, and subsequently Cohassack, from the Indian word Cooashauke, from cooash (pine) and auke (a place). The fall in this stream between the lake and river is one hundred and twenty feet, and a mill was built upon it as early as 1740 by John McMurphy.

Massabesic Lake is one of the finest natural features of Manchester. Massabesic is derived from massa (much), nipe (water), and auke (a place), "The place of much water," or the great pond place. It is nearly twenty-five miles around this lake, though nowhere much more than three miles broad. It is reputed to contain three hundred and sixty-five islands, and certainly does contain a great many, the largest of which is a seventy-acre lot, and is known as Brown's Island.

Indian names were always remarkably descriptive of a place, and the name "Massabesic," even as applied to this lake alone, is so; but it probably included and was applied to a section of territory extending northward, and including a pond beyond Rowe's Corner, so called, in Hooksett. If this is a correct view, no name could be more appropriate than "the place of much water." The Great Massabesic, the Island Pond and Little Massabesic, Clark's Pond, Swago or Swager's Pond, Tower Hill Pond, another at Rowe's Corner, and still another beyond the last mentioned and south of Bear Hill, all within a dozen miles, and most of them natural ponds. Taken together, this may be called by any man, red or white, most truly and emphatically the great pond place.

The large stream, which has its source in the north part of Hooksett, and flows, through each of the ponds named, into the Massabesic, is one of the feeders of that lake, and the time may come when, by clearing the forests along the upper waters of the Merrimack and its branches, it will become so diminished, that the great manufacturing companies on the lower Merrimack will feel the necessity of securing this great reservoir, and provide means to draw from it, as they have Winnipesaukee, Squam, and Newfound, to replenish the Merrimack when depleted by seasons of unusual drouth.

The Massabesic Lake extends into the town of Auburn, formerly a part of Chester. Auburn was incorporated in 1845. The people are generally engaged in agricultural pursuits and farming.

"Devil's Den Mountain" is an elevation near the eastern shore of the lake. At the base of this mountain is a large cave, the mouth or entrance being about three feet wide and six high. This subterranean passage extends nearly or quite to the centre of the mountain, opening into large apartments, several of which are fifteen feet square, and vary in height from one to five yards. Through an aperture too small to admit a person, other similar apartments are seen, with openings which are supposed to lead into others still beyond. The walls of this cave are of gneiss formation, and acrid to the taste.

This cave is called the "Devil's Den," and, though frequently explored, its extent is unknown; but it appears to be divided into several sections of apartments, and as his satanic majesty has not yet been seen in the neighborhood, or discovered in the cave, it is presumed his revels are kept in a remote and impenetrable section.

In some respects Londonderry is one of the most remarkable towns in New Hampshire, originally embracing a large territory from which in whole or in part several towns have been organized. It was incorporated in 1722, and has produced many eminent men. The settlement of this town was in 1719, and Parson McGregor preached the first sermon, from Isaiah xxxii. 2, under the shade of a giant oak, on the day after the arrival of the colonists.

McGregor, besides being a Christian minister, was a great admirer of the works of nature, and would go farther to see a curious tree, pond, or precipice than any other man in the colony. He was the first among them to visit Amoskeag Falls, having been attracted through fifteen miles of wilderness, with nothing to mark his course, by reports of their "hideous" grandeur.

Prior to this time the cultivation of the potato was unknown in this part of the country. Besides the introduction of the potato, this colony introduced the manufacture of linen.

Beaver River has its source in a pond of the same name in Londonderry, and falls into the Merrimack at Lowell.

There was a company from this town, consisting of seventy-five men, under the command of Capt. George Reid, in the battle of Bunker Hill, and as many more in the battle of Bennington, commanded by Capt. David McClary, who was killed in that battle.

The celebrated Dr. Thornton, on his arrival from Ireland, settled and practised medicine here.

Among the distinguished men born within the original limits of Londonderry, may be mentioned Gen. Stark, Col. Reid, Generals Miller and McNiel, — the two last distinguished officers in the last war with Great Britain, — J. McKean, first President of Bowdoin College, Judges Livermore, Bell, and Steele, Chief Justice Jeremiah Smith, and Attorney-General Prentice.

Derry, which was taken from Londonderry, is a wealthy town, and was the scene of the famous Derry Fairs. To "buy, sell, and exchange, and to indulge in sports, games, healthy and harmless recreation," was the design of these fairs; but in the course of time its doings becoming illegitimate they fell into disrepute and were discontinued.

It is said a certain man went to the fair on one occasion and called at a friend's house, hitching his horse at the gate. Finding the company jovial and agreeable he remained. During his stay in the house he received oft-repeated calls and invitations to "swap horses," which he uniformly assented to, with a one-dollar proviso, without leaving the house. When ready to return home, he was surprised to find the identical horse he came with tied to the post, while he had the snug little sum of thirty dollars in his pocket.

The Piscataquog River falls into the Merrimack on its west bank at Manchester. The sources of this river are in several towns, and the woods about its head-waters and branches supplied the Indians with large quantities of venison; hence its name. Pos (great), attuck (deer), auke (a place), meaning great deer place. The sources of this stream are high, consequently it is filled with falls and rapids. Mills were built upon it as early as 1775, at Piscataquog Village.

Francestown, where one of the sources of this stream is located, was named in honor of the wife of Gov. Wentworth. It was settled, by John Carson, in 1760, and incorporated twelve years later. There is a very valuable soapstone quarry here, the soapstone being used for sizing-rollers, stoves, and other purposes.

Weare supplies a large branch of the Piscataquog, and was formerly called Halestown. It has two societies of Friends or Quakers. The town was incorporated in 1764 under its present name, in honor of Meshech Weare, who was for some time "President" of New Hampshire.

Goffstown received its name in honor of the Goffe family. It was granted by the Masonian proprietors to Rev. Thomas Parker and others, of Dracut, Massachusetts, in 1748.

There are two mountains in this town very near together, bearing a strong resemblance to each other, and were some years signal stations for the United States Coast Survey. These mountains are called "Uncanoonucks," signifying the breasts of a woman. "This is a corruption of the Indian word Wunnunnuoogunash, the plural of Wunnunnoogun (a breast), ash being added to the singular to form the plural of inanimate names."

Amoskeag Village, at the Falls of that name, formerly belonged to Goffstown, but was annexed to Manchester in 1853. Rock Rimmon, near the village, is a prominent object, visible at a considerable distance. It is an outcropping of gneiss formation, being easily accessible on the west, while the south-east face is a perpendicular bluff some eighty feet high, its summit affording a splendid view of the river, the city of Manchester, and surrounding scenery.

Bedford is located on the west side of the Merrimack River, next below Manchester, and was formerly called Souhegan East. Piscataquog Village belonged to this town, but was annexed to Manchester in 1853. In this village is situated the residence of William Stark, Esq., who is not a pensioner on his distinguished ancestors.

He is a scholar, poet, and naturalist; and, being a gentleman of leisure and possessed of ample means, his zöological gardens contain the most extensive private collection to be found in New England, if not in the United States.

Bowman's Brook, so called from Jonas B. Bowman, Esq., some years deceased, has its course through the north-easterly portion of this town, and has been generally known until within a few years as the best trout brook in all this region.

John A. McGaw, Esq., a native of Bedford and a wealthy New York merchant and importer, has a fine summer residence near this stream, on the river road, one of the finest drives in the city of Manchester. Mr. McGaw's estate comprises about sixty acres of fertile interval; his buildings are elegant, and surrounded with fruit and ornamental trees. He has, also, one of the largest and best trout ponds in the State. The splendid location and the tasteful improvements give warrant to the expression, "altogether beautiful," as often applied to this situation.

Pulpit Brook has a fall of two hundred feet, and is a very picturesque cascade.

Bedford was granted, in 1733, to the officers and soldiers of the Narraganset War. It was settled four years after by Robert and James Walker, and, in 1738, Col. John Goffe, Matthew Patten, and Capt. Samuel Patten settled in the town. The first white child born here was Silas Barron.

On the bank of the river, near Goffe's Falls, an Indian burial-place was discovered, and, in 1821, Dr. Woodbury exhumed several skeletons, one of which was evidently buried in a sitting posture, and had long hair like a female, which was in a very perfect state of preservation.

Dr. Peter P. Woodbury, an eminent physician, practised medicine in this town until his death, about ten years since. He was a brother to the Hon. Levi Woodbury.

The Souhegan River, on which are located a number of mills and shops, empties into the Merrimack on its left bank. The Souhegans lived upon this river, occupying the rich intervals upon each bank above and below its mouth. Souhegan is a contraction of Souheganash, an Indian name in the plural number, meaning worn out lands. These Indians were often called Natacooks, or Nacook, from their occupying ground that was free from trees, or cleared land, — netecook meaning a clearing.

Ashburnham, in Worcester County, Massachusetts, was granted to Thomas Tileston and others, of Dorchester, in the year 1690, for services rendered in an expedition against Canada, and was called for many years "Dorchester Canada." This town is located on an elevated ridge, which is the water-shed between the Merrimack and Connecticut Rivers, and contains several large ponds, which are the sources of the Souhegan River.

Ashby, on the Souhegan River, is an enterprising and pleasant town in Middlesex County, Massachusetts, on the New Hampshire line, and contains mills, shops, and various mechanical works.

New Ipswich was granted by the Masonian proprietors in 1750, and was incorporated twelve years afterwards. It is one of the most enterprising among the back towns in the State, containing an academy, which was incorporated as early as 1789, and several cotton and other mills and shops. There were sixty-five men in the battle of Bunker Hill from this place. The first manufacture of cotton in New Hampshire was commenced here in 1803.

Milford, on the Souhegan River, was settled by John Burns and others. It was incorporated in 1794, and is a large and enterprising town. There are three hundred thousand dollars invested in cotton manufactures, the Souhegan Manufacturing Co. having been incorporated in 1846, the Milford Co. in 1810, and the "Pine Valley" Co. in 1867, capital one hundred thousand dollars. The Milford Plane Co., which produces the celebrated Eagle plane, is an extensive establishment, and employs from fifty to seventy-five hands. There are several other mechanical works in operation, and Milford is a busy, lively, and thrifty village.

Amherst, on the Souhegan River, was for many years the shire town of Hillsboro' County. It was first called Narragansett "Number Three," afterwards Souhegan West. The first settlement was in 1734, by Samuel Walton and Samuel Lampson. It was incorporated in 1760, under the name of Amherst. It is watered by Baboosuck River, which flows through two ponds, — Little and Great Baboosuck, — and, after leaving Amherst, passes through Merrimack, and unites with the Souhegan near the confluence of that river with the Merrimack.

Amherst has produced a large number of eminent men, among whom may be mentioned Hon. Moses Nichols, who was engaged in the battle of Bennington, being a colonel under Stark. He also held many important civil offices. Hon. Samuel Dana was a classmate of the elder Adams at old Harvard. He also held many positions of honor and trust. Hon. William Gordon was member of Congress in 1796, and afterwards Attorney-General of the State. Hon. Robert Means, a native of Ireland, settled in Amherst soon after its incorporation, and was elected to many high official stations. The Athertons — Joshua, Charles H., and Charles G. — were very eminent and distinguished men. Hon. Charles G. Atherton was Senator in Congress for several years, and was a prominent member of that august body, while at the bar it has been said he had but few equals. One of the most eminent members of the New Hampshire bar said of him : —

"In that gentleman are united many of the rarest qualifications of an advocate. Of inimitable self-possession, with a coolness and clearness of intellect which no sudden emergencies can disturb; with that confidence in his resources which nothing but native strength, aided by the most thorough training, can bestow ; with a felicity and fertility of illustration, the result alike of an exquisite natural taste and a cultivation of those studies which refine, while they strengthen the mind for forensic contests."

In 1853, Mr. Atherton was suddenly stricken in the court-house at Manchester. He was taken to his hotel, where he expired in a few days after the attack. His last words were : "I expected this, but not so soon."

Horace Greeley, of the "New York Tribune," is a native of Amherst, and, as a journalist, has probably achieved as much wealth, fame, and influence as any of its native citizens.

"Amherst jail," which was for so many years the terror of rogues, has been discontinued, — new and elegant county buildings having been erected at Manchester.

"The Amherst Cabinet" was first issued in 1802, and, though nearly threescore years and ten, is yet hale and vigorous.

Merrimack was formerly called "Souhegan East." It was settled in 1733, and incorporated in 1746. It claims the credit of making the first Leghorn bonnets, which often sold for forty or fifty dollars.

The Cromwell House, a trading-post, was erected here previous to its settlement, by John Cromwell, an Englishman; but, after indulging for some time in discreditable "tricks of trade," thereby enormously defrauding the Indians, his irregularities were discovered by them, and he fled precipitately, thus saving his scalp-lock, and leaving them to redress their grievances by burning the post.

There is at the present time a flourishing military school in Merrimack.

Litchfield, on the Merrimack, was formerly called Natticott. It was set off from the territory of Dunstable, and incorporated by Massachusetts in 1734, and was chartered by New Hampshire in 1749. This town contains only eight thousand five hundred acres, much of it being valuable timber land. It formerly belonged to and was occupied by the Souhegans, Naticook Island, in the Merrimack, now known as Reed's Island, being their summer residence.

The soil of Hudson being generally good, and it having the advantages of proximity to a good market, the land is in an excellent state of cultivation, and it is a thrifty farming town. The original grant of Dunstable included Hudson, which was incorporated in 1746 under the name of Nottingham West, which it retained until 1830, when its name was changed to Hudson.

Dunstable was a great resort of the Merrimack Valley Indians, and, after the settlement of Hudson, in 1710, they continued to come for a long time on peaceful trading expeditions from the headwaters of the Connecticut and the Merrimack. The fertile intervals of Hudson were cultivated by the Indians, who derived from them liberal supplies of maize, gourds, and beans. Here, as in Pennacook, they grew considerable quantities of corn, observing some regularity in planting, and being obliged to watch it constantly to prevent the crows — which they called "kawkont," from their peculiar scream — from destroying it.

They generally erected one or more wigwams in the cornfield, in which the children and others could remain and watch the field, keeping away the crows, but not killing them, as they were held as a kind of sacred bird. The Indians had a tradition that the crow brought their first corn and beans to them from their "Great Manit, Kautantonwits" field in the south-west.

The Nashua River, a beautiful tributary of the Merrimack, has its source in the Wachusett Mountain, in Massachusetts, and empties into the Merrimack, on its right bank, in the city of Nashua.

Like the Contoocook in New Hampshire, the fountain-heads which contribute to the formation of the Nashua are numerous, and located in several towns. Ashburnham, where the Souhegan takes its rise, is also one of the sources of the Nashua, while others are in Princeton, Holden, etc. The north and south branches unite in Lancaster, the latter being called Still River.

The Wachusett Mountain is three thousand feet high, and derives its name from a tribe of Indians inhabiting the territory about its base, and which, like the Nashuas, acknowledged the sway of the Pennacooks, and Passaconnaway as a sagamon or bashaba. Wachusett is derived from Wadchu (a mountain), and auke (a place).

"Thus the Wachusetts, in Indian vernacular, were the mountain place tribe. The Indians considered the country along the whole course of the Nashua, from its head-waters to its confluence with the Merrimack, as one of the most delightful sections of this country, and, it is said, the early white settlers also regarded it as a pleasant country and a most superb stream, fully realizing its picturesque scenery before the demands of civilization and the hand of art had marred its primitive and charming natural beauty."

Wachusett Mountain and its vicinity continue to be a great resort in the summer season for comfort and pleasure-seekers, and may be sought by all admirers of nature and of charming natural scenery.

The Nashuas occupied the lands bordering that stream to its mouth, and the intervals on either side of the Merrimack. The tribe took its name from the river, and the latter derived its name from the Indian word which signifies "the beautiful stream with a pebbly bottom."

Lancaster, through which flows the Nashua River, is one of the finest inland towns in New England. The eye of the stranger is attracted and his attention arrested by the variety of its scenery and the natural beauty and pleasantness of the place. For many years it was on the frontier of civilization, and is one of the oldest towns in this section, it having been settled as early as 1643, and incorporated ten years later, and for years after this period was exposed to the predatory and hostile incursions of the Indians. In 1676, an attack was made by a large body of them, some fifteen hundred in number, upon the town. After a desperate resistance, many having been slain on both sides, the village was destroyed, and several persons taken captive, among whom was the celebrated Mrs. Mary Rowlandson.

Lancaster contains a large number and variety of mechanical shops, cotton and woollen mills, and others where hats, shoes, musical instruments, metallic and wooden utensils, and furniture are made. Lancaster gingham is widely famed as a superior fabric.

Harvard, on the Nashua River, contains a community of the sect called Shakers. This town was taken from Groton, Lancaster, and Stow, and incorporated in 1732. There is a slate-stone quarry, from which gravestones and monuments are manufactured in large numbers. The town also produces quantities of paper, shoes, and hats.

The Nashua River provides the town of Shirley with excellent water-power, which is improved, there being several cotton and woollen mills in operation. Shirley is an uneven township; but the soil is generally strong and good, while the large tracts of interval along the Nashua are fertile and productive.

Groton was incorporated 1665, and in 1676 was pillaged and destroyed by the Indians, the inhabitants taking refuge in garrison-houses provided for that purpose. Groton is a wealthy town. The facilities for education are excellent, there being several institutions of learning. But it is best known as the grand central point or junction of several railroads, by which it has become a great thoroughfare of travel.

Pepperell was incorporated 1753. The Nashua River supplies good mill privileges, and paper mills and shops for various mechanical purposes are in operation. The town was named in honor of Sir William Pepperell, who was chosen, in 1727, one of His Majesty's Council, which office he retained until his death. He was commander of the expedition against Louisburg, which was successful in all respects, and was accompanied by Dr. Thornton, of Londonderry, N. H., better known as Matthew Thornton, one of the signers of the Declaration of Independence. For his brilliant success in this expedition the king honored him with the dignity of baronet. He died in Kittery, Maine, where he resided, July 6th, 1759, aged sixty-three years.

William Buttrick, Esq., an enterprising and active business man, who owned and conducted the paper mills at this place, contributed much to the prosperity of the town. Mr. Buttrick died about twenty-three years since.

Hollis was settled in 1731 by Peter Powers; and his son, also named Peter, was the first white child born in the town. It was incorporated in 1746, and named in honor of Hollis, Duke of Newcastle. It was formerly called Nisitissit, and was afterwards known as Dunstable West Parish. It is watered by the Nashua and Nisitissit, one of its branches.

Nashua is historic ground. The scarcely tangible shadows of uncertain tradition have blended and mingled with the substance of her genuine, undoubted, and splendid records until the very truthfulness of details produces what may be properly styled the romance of history. Too fearfully and painfully true, on the one hand, for mere romance, while the romantic chivalry, wild adventure, and deeds of daring, trials, sufferings, and successes, the sturdy and conspicuous valor of her sons, who won unfading glory in fierce, yet most unequal conflicts, give a meaning and a significance to, and illustrate the adage that "truth is stranger than fiction."

Coming as they did to this spot by the beautiful river so beloved by the Indians, to their hearthstones and tillage fields, it is not strange the early settlers should have been pursued by them with implacable and vindictive animosity, and the woods, rocks, and hills concealed a lurking, deadly foe. Personal safety was never counted on, and he who was obliged to go forth on business, or to his daily labor, took his life in his hand, and went at its peril.

On one occasion two men, named Cross and Blanchard, were at work in the woods, when they were suddenly set upon by a party of Indians and captured. Not returning at night, suspicion of their probable fate was aroused, and a party of ten resolute and well-armed men, under the command of Ebenezer French, started in search of them. On arriving at the place where they had been at work it was at once discovered that Indians had been there and captured them; but the party learned by signs purposely left by them that they were living, and the indications were positive that the savages had but recently fled. Under these circumstances immediate pursuit was, without hesitation, determined upon. At this point of the proceedings, Josiah Farwell, who was one of the party, suggested, as a precautionary measure, that they should take a circuitous route and move warily; but it appears that some difficulty existed between him and the leader, French, when the latter, sneeringly, exclaimed, "I am going to take the direct path; if any of you are not afraid, let him follow me." * French led the party, as he declared he should, straight up along the Merrimack, and, having reached the vicinity of the place now known as Thornton's Ferry, they were fired upon and most of them killed. All the others, except Farwell, soon met the same fate. He alone escaped, and, collecting another strong force, returned and secured the bodies of eight of the victims, and had them decently buried. Their names were Lieut. Ebenezer French, Thomas Lund, and Oliver Farwell, of Dunstable; Daniel Balding and John Burbank, of Woburn; Mr. Johnson, of Plainfield; and one other not known.

The following year, when Lovewell made his expedition against the Pequaukets, he selected Josiah Farwell as his lieutenant, on account of his well-known prudence and skill in Indian fighting. When Lovewell's company had been recruited and organized by the selection of himself, Farwell, and Robbins, these officers were requested to petition the government for encouragement. Accordingly the following petition was sent by them to the General Court of Massachusetts: —

"The humble memorial of John Lovewell, Josiah Farwell, Jonathan Robbins, all of Dunstable, showeth: —

"That your petitioners, with nearly forty or fifty others, are inclinable to range and to keep out in the woods for several months together, in order to kill and destroy their enemy, Indians, provided they can meet with encouragement suitable. And your petitioners are employed and desired by many others, humbly to propose and submit to your Honors' considerations, that if such soldiers may be allowed five shillings per day in case they kill an enemy and possess their scalp, they will employ themselves in Indian hunting one whole year; and, if within that time they do not kill any, they are content to be allowed nothing for their wages, time, and trouble.

"JOHN LOVEWELL.
"JOSIAH FARWELL.
"JONATHAN ROBBINS."

"*Dunstable, November,* 1724."

In reply to the petition, the objects which it set forth were ap-

---
* Belknap.

proved, and a bounty of one hundred pounds was voted for "each Indian scalp taken by Lovewell's company."

The terrible experience of the people of Dunstable was, however, only a magnified transcript of the sufferings of other frontier settlements, but the record of her trials and the history of her glorious deeds cannot fail to bring to mind the turbulent days of unmitigated hostility, and the incessant alarms of exterminating war, —

> "What time the noble Lovewell came
> With fifty men from Dunstable,
> The cruel Pequot tribe to tame,
> With arms and bloodshed terrible," —

and enrolls a long list of her brave and gallant sons for heroic and self-sacrificing devotion among the bold, the strong, and the daring of their countrymen who have in arms achieved enduring historical renown and lasting gratitude.

The city of Nashua was formerly embraced within that large tract of country which was chartered in 1673 under the name of Dunstable. This tract of land included all the territory of the towns of Dunstable, Tyngsboro', and portions of Townsend, Groton, Pepperell, and others, in Massachusetts, and Hollis, Brookline, Milford, Amherst, Merrimack, Hudson, Litchfield, and parts of the territory of other towns in New Hampshire, and was under the jurisdiction of Massachusetts until the State line was permanently established in 1641. It was incorporated by New Hampshire, in 1746, under the same name, which it retained for ninety years after.

In 1803, the first post-office was established at Nashua, the name of the place was changed to Indian Head, and a scow was launched with ceremony and christened "The Nashua," and the proceedings were marked with great parade and display. At that time an "inn," a store, and three or four houses were the sum total of the present large and flourishing city of Nashua. The population in 1820 numbered only one thousand one hundred and forty-two.

In 1842, there was a disagreement about the location of the town house. It was located on the south side of the Nashua River, and the people on both sides claimed it. A great deal of feeling was manifested on both sides; the dispute waxed warm, and the town was divided. The north side, and a small tract of the north-east part of the south side, were incorporated by the name of Nashville. This state of things existed until 1853, when a city charter was obtained, under which the divided and estranged family was brought together. The "House of Nashua" — divided against itself — was reunited, Nashville was mollified, the "fatted calf" was killed, and, shoulder to shoulder, hand in hand, they have advanced through these intervening fifteen years until now Nashua is one of the principal cities in wealth, population, beauty, enterprise, and prosperity of the old Granite State.

Cities, like animals, thrive and increase, fatten and flourish, on the sustenance with which they are supplied. Thus, New York gained rapidly on commerce; St. Louis made a healthy beginning on buffalo robes, and peltries generally; New Bedford does well on whales; Newport, on the folly of shoddy nabobs; Holton, on copper mines; Pittsburg, on iron and coal; Gloucester, on fish; Chicago, on wickedness; Bangor, on lumber; Lynn, on shoes; and Nashua on manufactures; and her growth has been rapid and substantial.

In 1823, the water-power and adjacent lands were secured for the purpose of improvement, and, before the end of three years, the Nashua and the Jackson were in operation, and before the end of twenty-five years the city numbered more thousands of population than it did hundreds at the time these companies commenced operations.

"The Pennichuck Water Works were constructed in 1854. The Pennichuck has its rise in a pond near the north-western boundary of the city, and is fed by many never-failing springs of soft, pure water, and falls into the Merrimack. The water is taken at a point just above the Concord Railroad, from an artificial pond of twenty-six acres, and forced by a turbine wheel of eighty horse-power into a reservoir, half a mile north of the City Hall, one hundred and ten feet above the street level at that point, and of a capacity of one million two hundred and fifty thousand gallons."

The Nashua Manufacturing Company was incorporated in June, 1823, with a capital of one million dollars.

Salmon Brook, which rises in Groton, Massachusetts, passes through Nashua, where mechanical shops are located along its course. In fact, the city seems one extensive workshop, where mechanics and artisans congregate and flourish, and where every variety of machinery and implements is turned out, — massive and ponderous, delicate and curious, useful and ornamental, — which, in perfection, diversity, quality, and value, in detail and in the aggregate, are claimed to be unsurpassed by few cities of its size in New England.

There are three newspapers published here, and the facilities for good common-school education are not inferior to any other wide-awake place : a commendable public spirit, combined with experience, having provided suitable buildings, teachers, and all the necessary appliances for the most approved and thorough system of public schools.

Living as they did in an era of stirring events, many of the first settlers of Nashua became historic characters, and are still represented among the prominent and most influential of its citizens. The tombstones in the old cemetery below the city (the oldest burial-place west of Portsmouth and Hampton) present a long list of names well known to fame ; among them that of John Lovewell (father of the hero of Pequauket), who died in 1754, aged one hundred and twenty years. Among those who aided materially in bringing the city of Nashua into existence, Gen. Noah Lovewell was the first postmaster (1803); Daniel Abbot, Esq., was a prominent lawyer and active business man, and B. F. French was, at one time, his law-partner, afterwards agent of the Boott, at Lowell, and other corporations; Israel Hunt was an officer in the battle of Bunker Hill; his sons, Gen. Israel and John M. Hunt, Esq., are still among the foremost men of Nashua. The Frenches, Benjamin and Thomas, were also among the first families, having descended, on the mother's side, from the celebrated Jonathan Blanchard. Col. Joseph Greeley, another of the old stock, died but a few years since. Hon. Amos Kendall, postmaster-general under Gen. Jackson, was also one of the prominent men who originated in old Dunstable.

Among the foremost business men of Nashua is Gen. George Stark, a descendant of the famous Stark family, of Derryfield, and brother to William Stark, the poet. Gen. Stark is one of the leading railroad men of New England, and has been a candidate for Governor of New Hampshire, a position he is eminently qualified to adorn. Col. Thomas P. Pierce went to Manchester to reside in 1840; in 1847 he enlisted for the Mexican War, and was appointed lieutenant in the regular army. He served through the war with distinction, and was complimented for gallant and meritorious conduct. In 1853 he was appointed postmaster at Manchester by President Pierce, and again by President Buchanan, and the people of that city endorsed his administration of the duties of his position for eight years with entire unanimity, and, to this day, he is referred to as the "model postmaster." In 1861 he was appointed colonel of the 2d New Hampshire Regiment, but was, however, in consequence of the derangement of his private business, compelled to resign before taking the field. A few years since Col. Pierce removed to Nashua, where he is counted among the foremost of the younger portion of her active and enterprising business men.

In addition to the Nashua and Jackson Companies, the Iron Company has a capital of one hundred and twenty-five thousand dollars; the Nashua Lock Company, a capital of sixty thousand dollars; Underhill Edge Tool Company, a capital of eighty thousand dollars; and the Francestown Soapstone Company, a capital of three hundred thousand dollars.

The fine natural features of Nashua supply her with the elements of a beautiful city. The pellucid Nashua meandering through its very centre, her fine streets, splendid buildings, and grand old ornamental trees, public buildings, churches, factories, and workshops, the hum of the spindle, the ring of the anvil, the constant rumble of car-wheels, all combine to give the city an air of beauty, enterprise, and thrift most attractive and agreeable.

Tyngsboro' was formerly a portion of the Dunstable charter, and was incorporated, 1789, under its present name, which it received in honor of the Tyngs, who were famous Indian fighters among the celebrated names in old Dunstable.

In 1734 Massachusetts granted a tract of territory, bounded on the west by the Merrimack, and on the east by a line three miles from that river, extending from the north line of Litchfield to the Suncook River, which included portions of what is now Londonderry, Manchester, and Hooksett, to Col. William Tyng and others, as a reward for their services in fighting the Indians. This service consisted of raising a company of the daring and prudent men who penetrated the enemy's country as far as the Winnipesaukee, on snow-shoes, in the depth of winter, killing six Indians, and dispersing all the bands with which they came in contact. This grant was called Tyngstown, but, as Massachusetts had only the right of the usurper to make this grant, it was, of course, void and of no effect.

On hearing of the disaster to Capt. Lovewell's forces at Pequauket, Gov. Dummer forthwith despatched a company under Col. Eleazer Tyng to succor the living, if possible, and recover the dead. Col. Tyng marched on the 17th of May, and on the following day wrote Gov. Dummer : —

"MAY IT PLEASE YOUR HONOR: —

"This day I marched from Amoskeag, having fifty-five of my own men and thirty-two of Capt. White's. The men are well, and proceeded with a great deal of life and courage. Yesterday I was forced to lie still, by reason of the rain. I would humbly offer something to your Honor in the behalf of our people who are left destitute and naked, that you would be pleased to consider their circumstances, and order what you shall think proper for their defence till we return. "I am your Honor's
"Most ob't servant,
"ELEAZER TYNG.

"*Amoskeag, May 19th, 1725.*"

Gov. Dummer issued the following order : —

"TO COL. FLAGG: —

"Sir, — These are to empower and direct you forthwith to detach or impress out of the regiment where you are lieut. col., a sergeant and twelve effective, able-bodied men, well armed for his Majesty's service, for the security and reinforcement of Dunstable, until the return of Col. Tyng and his company.

"They must be posted at the garrisons of Joseph Bloghead, Nathaniel Hill, John Taylour, and John Lovewell, and three sentinels in each garrison, and the sergeant in that of the four that is nearest the centre.

"The sergeant must be very careful to keep the men well upon their duty, so as to be a good guard and protection to the people, and you must give them directions in writing accordingly. Let the matter be effected with all possible despatch.
"WILLIAM DUMMER.

"*Boston, May 19th, 1725.*"

The force under Col. Tyng was successful in finding the battle-ground, but discovered no Indians. They found and identified the bodies of Capt. John Lovewell,[*] Ensign Jonathan Woods, Ensign

[*] "With footsteps slow shall travellers go,
  Where *Lovewell's* pond shines clear and bright,
  And mark the place where those are laid
  Who fell in *Lovewell's* bloody fight."

This "poem," descriptive of Lovewell's famous fight, was published the year after the battle, and republished in the N. H. Hist. Soc. Pub., Vol. III.

John Harwood, and Robert Usher, all of Dunstable; Jacob Fullam, of Weston; Jacob Farrar and Josiah Davis, of Concord; Thomas Woods, Daniel Woods, and John Jefts, of Groton; Ichabod Johnson, of Woburn; Jonathan Kittredge, of Billerica."

Col. Tyng also found an Indian grave, which he opened. It contained among other bodies that of the dreaded Paugus.*

The scene of this battle, where the intrepid Lovewell and nearly all his force were slain, was in the "Pequaquauke country." These Indians, made up of the remnants of the various tribes or bands of the Pennacook confederacy, had located themselves upon the branches of the Saco, where was an abundance of fish and game, and, from the fact that their main village was upon the Saco, near where that river makes a noted bend or circuit of some thirty-six miles, principally in Fryeburg, Maine, returning within a mile or two of the Indian village where it commenced its detour, were called Pequaquaukes, or Indians at the crooked place, — Pequaquaukes being derived from the adjective *pequauquis* (crooked) and auke (a place).

This name, thus received, was applied to the Indians of all that region of country, and has ever since been applied to the region of country itself. The Pequaquaukes were under the control of two powerful sagamons, Paugus (the oak) and Wahowah (the broad shouldered).

The Tyng family of Dunstable appear to have been leading and prominent people for many years among the illustrious names of that distinguished settlement. Previous to 1700, when a small grant was made to Wonnalancet, that mild and amiable chieftain was placed under the charge of Mr. Jonathan Tyng, of Tyngsboro'.

This town is situated on both sides of the Merrimack, which is here a broad and beautiful river. Having pursued a general southerly course since leaving old Lafayette, the Merrimack commences a grand detour at this place, maintaining a remarkably regular trend, and in the course of fifteen miles heads due east, and continues in the same direction to the sea.

---

* "'Twas *Paugus* led the Peqw'k't tribe;
    As runs the fox, would *Paugus* run;
    As howls the wild wolf, would he howl;
    A huge bear-skin had Paugus on."

Chelmsford is a very ancient town, having been settled more than two hundred years ago. For many years the product of manufactured articles has been in the aggregate large, consisting principally of glass, iron, cotton, and wool fabrics.

Stony Brook, which rises in Groton, unites with the Merrimack in this town. On this stream are many mills, and quite a large investment of capital in various kinds of manufacturing. Among the largest of these may be mentioned the Abbott Worsted Company, Graniteville, capital seventy-five thousand dollars; Nail Factory at Forge Village, capital thirty thousand dollars; Eagle Mills, West Chelmsford. At North Chelmsford are located the Swain Turbine Company, with a capital of fifty thousand dollars; the Baldwin Company (worsted yarn), capital fifty-two thousand dollars; Foundry Company, capital sixteen thousand dollars, and the Sheldon Hosiery Company, with a capital of one hundred thousand dollars.

Previous to the charter of Dunstable, Chelmsford was an outpost of civilization, and subsequent to that time it appears that a garrison was maintained, and if cruelty was practised on occasions by the Indians, it was also perpetrated on them in turn without cause, and hardly less savage than their own.

Sometime about 1680 the barn of Lieut. James Richardson was destroyed by fire, and its destruction was charged to the Indians, and fourteen men, armed with guns loaded with buckshot, repaired to wigwams of the Wamesits, called them out, and two of the men discharged their pieces among them, killing one boy and wounding five of the women and children. It is true that the men, Lorgin and Robins, who had fired upon them, were tried; but of course, such was the feeling against the Indians, they were acquitted, "to the great grief and trouble generally of magistracy and ministry and other wise and godly men."

This was in December, and in the February following the Wamesits sent a petition to the governor and council, requesting to be removed from Chelmsford, fearing that hostile Indians would make reprisals for this cowardly act, alleging that in such an event they would again suffer without cause; but the governor paid no attention to their request, and they fled into the wilderness towards Pennacook, leaving behind in one wigwam several sick, lame, and blind. After they had retired, the people of Chelmsford assembled and set fire to this wigwâm, which, together with these poor Indians, six in number, was destroyed.

At the time of the settlement of Chelmsford, the beautiful island in the river here, which was called Wickasauke, was a spot of considerable consequence, and no little interest. It appears that this island belonged to the family of Passaconaway, and was an important cornfield, and the occasional residence of his successor, Wonnalancet. The elder brother of the latter, having become surety for another Indian, was taken and lodged in jail in Boston. The kindhearted Wonnalancet set about providing means to release his unfortunate brother, and all else failing he sent the following petition, requesting to be allowed to sell the island that he might release Nanamocomuck from prison. This petition received the following answer: —

"*License for Indians to sell an Island.* — Whereas this Court is Informed yt Peasconaway's soune now in prison as surety for ye payment of a debt of forty-five pounds or thereabouts, and having nothing to pay but affirms that several Indians now in possession of a smale island in Merrimack River (about sixty acres), the half whereof is broken up, are willing after this next yeares use of their sayd island to sell theire interest in ye sayd island to whoever will purchase it, and so redeem the sayd Peasconaway's soune out of prison. The Magistrates are willing to allow the sayd Indians liberty to sell ye sayd island to Ensigne Jno. Evered as they and he can agree for ye ends aforesaid. If their brethren the deputys consent hereto. 8 Nov. 1659. The deputys consent hereto provided the Indians have liberty to sell the said island to him that will give most for it.

"Consented to by ye Magistrates.
"EDWD. RAWSON, *Sec'y.*"

The island was sold. Ensign John Evered, or Webb as he was sometimes called, being the purchaser, and Nanamocomuck was released from jail. Wonnalancet, however, received a grant, through pity, of one hundred acres, which was located on a hill ten or twelve miles west of Wickasauke. In 1665 another petition came from him concerning the island: —

"To the Worshipful Richard Bellingham, Esq., Gov'r, and to the rest of the Honord Generall Coart.

"The petition of us poore nelbour Indians, whose names are hereunto subscribed, humbly sheweth that whereas Indians severall years since we yr petit's out of pity and compassion to our pore brother and countryman to redeem him out of prison and bondage whose name was Nanamocomuck, the elder son of Passaconewa, who was cast into prison for a debt of another Indian unto John Tinker, for which he gave his word; the redemption of whome did cost us our desirable posetions where we and oure had and did hope to enjoy our livelihood for ourselves and posterity; namely, an island on Merrimack River, called by the name of Wicosurke, which was purchased by Mr. John Web, who hath curtiously given vs leaue to plant vpon ever since he hath possessed the same, we doe not know wheather to goe, nor where to place ourselves for our Lively hood in procuring vs bread; having beine very solicitious wh Mr. Web to let vs enjoy our said posetions againe, he did condescend to our motion provided we would repay him his charges, but we are pore and canot so doe. Our request is Mr. Web may have a grant of about 5 C acres of land in two places adjoyning his owne lands in the wilderness, which is our owne proper lands as the aforesaid island ever was.

" 10; 8; 65. Nobhow in behalf of my wife and children,

" VNANUNQUOSETT.
" WANALANCETT.
" NONATOMENUT.

" If the court please to grant this then yr petitioner Wanalancett is willing to surrender up ye hundred acres of land yt was granted him by the court."

The reply to this petition was favorable, and as follows : —

" In answer to this petition the Court grant Mr. Jno. Evered five hundred acres of land, upon condition hee release his right in an island in Merimacke River, called Wicosacke, which was purchased by him of the Indian petitioners; also upon condition Wonalancett do release a former grant to him of an hundred acres, and the court do grant said island to petitioner. John Parker and Jonathan Danforth are appointed to lay out this grant of five hundred acres to John Evered.

" EDWARD RAWSON, *Secretary.*
" Consented to by the Deputies.

" 14 October, 1665."

Wonnalancet and his friends continued to occupy and cultivate this island for some years, under the supervision of Mr. Jonathan Tyng, of Dunstable, and Mr. Robert Parris, his " bayl." Yet this noble child of the forest, after his experience, having seen the practical workings of a faith, which he was importuned to adopt, in the uncalled-for and inhuman slaughter of his defenceless people at Chelmsford, distrustful, if not of Christianity, certainly and with good reason of its professors, was himself a practical and genuine Christian.* Having been crowded out of possession by the grasp-

* He was always friendly to the English, but unwilling to be importuned about adopting their religion. When he had got to be very old, however, he submitted to their desires in that respect. Upon that occasion he is reported to have said, *I must acknowledge I have all my days been used to pass in an old canoe, and now you exhort me to change and leave my old*

ing and constantly encroaching English, he finally retired to the St. Francis.

As a proof of his great but unobtrusive friendship towards the English, it is related of him that he came to Chelmsford after an absence of some years, and meeting the Rev. Mr. Fisk asked him the news, and if Chelmsford had suffered during the Indian War. "No," said Mr. Fisk, "it has not, thank God!" "Me next," replied Wonnalancet, who had evidently interposed his personal, friendly feeling, his influence, and his authority to shield Chelmsford from the doom which perhaps otherwise impended over it. Another instance of his unwavering friendship and Christian spirit may be given. Wonnalancet informed Major Hinchman, commander of the post at Chelmsford, that the Mohokes were in the valley of the Merrimack, above Dunstable, and in the neighborhood of the mouth of the Souhegan River. James Parker at once communicated this information to the governor and council in these terms: —

"From Mr. Henchman's farme ner Meremack, last post hast.

"To the Honered Govner and Counsell. This may informe youer honores that Sagamore Evanalanset (Wonnalancet) came this morning to informe me, and then went to Mr. Tyng's to informe him, that his son being on ye outher sid of Meremack River a hunting, and his dauter with him, up the river, over against Souhegan, upon the 22d day of this instant, about ten of the clock in the morning, he discovered 15 Indens on this sid of the river which he soposed to be Mohokes by their speech, and he having a canow ther in the river, he went to breck his canow that they might not have ani ues of it, in the menetime thay shot about thirty guns at him, and he being much frighted, fled and came home forthwith to Nahamcok, wher ther wigowemes now stand.

"Not eles at present, but remain your servant to command,

"JAMES PARKER.

"Rec'd 9 night 24 Mrh. 76-7."

As the rolling seasons each in turn present a charm peculiarly its own, differing in degree less than in kind, each indiscriminately attractive to those who properly realize and appreciate the wonderful economy and perfection of nature; so likewise does this beautiful

canoe and embark in a new one, to which I have hitherto been unwilling, but now I yield up myself to your advice, and enter into a new canoe, and do engage to pray to God hereafter.—(Drake's Book of the Indians.)

The Apostle John Eliot wrote to Hon. Robert Boyle, in England, 1677, as follows: —

"We had a sachem of the greatest blood in the country submitted to pray to God a little before the wars; his name is Wanalauncet."

and bright rolling river, ever varying, ever charming, dashing down the mountain side in countless noisy rivulets, each directing its course, not perhaps in a right line, but with ultimate certainty to the same great channel, as the races of men, sweeping down from the misty summits of antiquity and barbarism, are destined finally to unite in one homogeneous mass, when brotherly love, "peace on earth," and good will shall be the prevailing elements of the mighty but tranquil current, then meandering through the secluded dells beneath the shade of great woods whose shadows have danced for long ages upon its buoyant ripples, or dashing in foam and fury over interposing rocks and falls, narrowing to deep and quiet channels, again expanding, as at Chelmsford, where it spreads to the dimension of a broad and majestic river.

The Pawtucket dam has the effect to set the water back far above Chelmsford, and to give the river a channel deep, wide, and still, and for miles its placid and polished surface and its graceful curve are very marked and noticeable features of the natural scenery of this vicinity. When the great lumber "drive" from Woodstock and the ungranted lands reaches here, and is shut into the river by the Lowell boom, what can be more picturesque than to see ten million feet of round lumber floating on its surface. This sight, though differing from the "hideous" falls and many other famous points in the course of this river, is grand indeed. Logs and lumber, in all forms, were originally "rafted" down this river in the following manner: The lumber was ranged in convenient lengths, side by side, until it reached a breadth to navigate handily, and then securely fastened. The credit of introducing the "drive" system on this river, which is more economical and rapid, and a decided improvement in all respects, is due to Nicholas Norcross, Esq.

Dracut was incorporated in 1701. Beaver River, which rises in Londonderry, New Hampshire, falls into the Merrimack on its left bank in Dracut; and on this stream, about one mile from its mouth, are extensive woollen manufactories, at a place which has long been familiarly known as Dracut Navy Yard. This river, like most other tributaries of the Merrimack, affords many mill privileges, which are well employed, either for extensive manufactories, or for the saw and grain mill and other works which accommodate the local business and supply the needs of the people living along its course.

But a mile or so from its source there was, in Derry, on this stream, a small mill, and in the village is a grain and saw mill and mechanical works, also a clothing-mill; at Londonderry a grain-mill was erected sixty-five years ago; a saw-mill and small factory, also Butler's saw and grain mill; in Pelham there is a privilege which was first operated sixty years ago; the Ames grist-mill and saw-mill in Dracut, now owned by the Merrimack Woollen Mills Company, and used as a woollen-mill; the Goodhue privilege, owned and operated by Charles Richmond for making paper; at the "Navy Yard," so called, is the Merrimack Woollen Mills Company. As early as 1815 this privilege was owned by the Messrs. Stanley, who manufactured cotton yarns. Since that time several parties have operated this power for the manufacture of satinets, flannels, blankets, etc., and about thirty years since a Mr. Aiken made patent saw sets and awls in the old mill. The Merrimack Company erected, four years since, a large brick mill, where they make fancy cassimeres; capital five hundred thousand dollars.

Among the foremost of those men whose career is intimately connected with the Merrimack River, and who have wrought for themselves an ample fortune and a well-merited fame, is John Nesmith, Esq., a native of Londonderry, N. H., and for many years a resident and one of the leading citizens of Lowell, who is a large owner in this company. There is scarcely an enterprise of any magnitude along the Merrimack, from the great lake to tide water, but is due for its success, often even its inception, to the sagacity of Mr. Nesmith. It is more than probable that he is entitled to the credit of founding the great manufacturing city of Lawrence. As early as 1836, Mr. Nesmith, in company with Daniel Saunders, Esq., had made purchases of the land adjacent to the falls on either side of the river, and had secured a charter for damming. The financial revulsion of the following year, however, checked the progress of this enterprise, and it was not until 1844 that the scheme of Mr. Nesmith was consummated. About this time he secured a renewal of his former charter, when the heavy capitalists of Boston were induced to engage in business here, and the result is well known. Mr. Nesmith formerly owned a mill at Hooksett, N. H., which he sold to the Amoskeag Company, and at the time of the origin of the latter company and of Manchester, he was employed to purchase lands, and secured the first lot bought by them on the east side of the river. To his view the "Moore Farm" was the key to operations at Manchester; and this he secured, his judgment proving of great importance to complete success. It was Mr. Nesmith who first conceived the project of securing rights on Lake Winnipesaukee, and the feasibility of providing artificial means to draw from it in dry seasons. When he suggested the idea to the manufacturers along the river, it met with so little attention that he, fully appreciating the value of such a great reservoir, went on and bought the right to control the lake on his own account, and it was only a short time before the companies were anxious to purchase Mr. Nesmith's claim at a handsome premium, when they completed the means of reducing the lake several feet in the event of a short supply of water occurring, which was exactly Mr. Nesmith's idea. In 1831, Mr. Nesmith made Lowell his permanent residence, purchasing the Livermore estate (about one hundred and thirty acres), which he laid out and sold for building-lots, realizing thereon a gratifying advance. Mr. Nesmith has never known, or sought to know, the sensations or emotions of idleness, and for the past forty years has been constantly and actively engaged in manufacturing operations, — in cotton, worsted, flax, printing cloths, etc., etc., nearly through the list of textiles. For many years he has been an extensive owner of stock in the Merrimack Company, as well as in other corporations in Lowell and Lawrence, banks, shipping, etc. Mr. Nesmith has made valuable contributions, both in money and intellect, to the perfecting of improvements in machinery, some of them of great importance, and has also been ever prompt with his ample means in advancing any object of interest that would tend to enhance the growth or prosperity of the city of his adoption, — ready to contribute to all charitable and worthy purposes the full and ample measure of his bounty. Although an octogenarian, he is still clear-headed, vigorous, and masculine in intellect, and retains a remarkable physical elasticity. Never having had time for that purpose, he has kept aloof from active politics, except when called by the voice of his fellow-citizens, to which he has felt it his duty to respond. He has held many responsible positions, among others the office of lieutenant-governor, and is at the present time Collector of Internal Revenue for the Lowell District.

CHAPTER X.

Pawtucket Falls. — Indians. — Canals. — Lowell. — History of Manufactures on the Merrimack. — The Concord River. — Its History. — History of its Manufactures.

"The Indians in this neighborhood were sometimes called Pawtuckets, from the falls in the Merrimack of that name. Pawtucket means the forks, being derived from the Indian word pohchatuk (a branch). Pawtucket seems, however, to have been applied by the English rather to all the Indians north of the Merrimack, than to the particular tribe at the falls of that name."

When the constantly increasing tide of population had rolled back from the original sea-girt colony, they found at the Pawtucket Falls, in the town which was afterwards called Chelmsford, the head-quarters of a tribe of the confederated Pennacooks, with Passaconnaway as the sagamon. Then, as now, it contained a large community, was a place of great note and importance, supplied the community with sustenance; then, as now, derived its importance solely and entirely from these falls. This was one of the three great principal fishing-places on the Merrimack, belonging exclusively to the confederacy which was under the rule of a wise and sagacious chief.

In 1660, Passaconnaway, having reached the venerable age of more than fourscore years, appointed a day for the assembling of all the subordinate chiefs and principal men of the tribes of the Pennacooks, or Pawtuckets, as they were generally called by the English, for the purpose, voluntarily and deliberately formed, of abdicating in favor of his son Wonnalancet, which he did in the following speech, sometimes called his "dying speech;" but, as he was alive some years after, it may be more properly called his "farewell address" to his people: —

"Hearken," said he, "to the words of your father. I am an old oak that has withstood the storms of more than one hundred winters. Leaves and branches have been stripped from me by winds and frosts;

my eyes are dim, my limbs totter, — I must soon fall! But when young and sturdy; when my bow, — no young man of the Pennacooks could bend it, — when my arrows would pierce a deer at a hundred yards, and I could bury my hatchet in a sapling to the eye; no wigwam had so many furs, no pole so many scalp-locks as Passaconnaway's! Then, I delighted in war. The whoop of the Pennacooks was heard upon the Mohawk, and no voice so loud as Passaconnaway's! The scalps upon the pole in my wigwam told the story of Mohawk suffering! The English came; they seized our lands; I sat me down at Pennacook. They followed upon my footsteps. I made war upon them, but they fought with fire and thunder; my young men were swept down before me when no one was near them. I tried sorcery against them; but still they increased, and prevailed over me and mine, and I gave place to them and retired to my beautiful island of Natticook. I that can make the dry leaf turn green and live again, — I, that can take the rattlesnake in my palm as I would a worm, without harm, — I, who have had communication with the Great Spirit, dreaming and awake, — I am powerless before the pale-faces. The oak will soon break before the whirlwind; it shivers and shakes even now; soon its trunk will be prostrate; the ant and the worm will sport upon it! Then think, my children, of what I say. I commune with the Great Spirit. He whispers me now: —

"'Tell your people, peace, peace is the only hope of your race. · I have given fire and thunder to the pale-faces for weapons. I have made them plentier than the leaves of the forest, and still shall they increase! These meadows they shall turn with the plough; these forests shall fall by the axe; the pale-faces shall live upon your hunting-grounds, and make their villages upon your fishing-places.' The Great Spirit says this, and it must be so! We are few and powerless before them! We must bend before the storm! The wind blows hard! The old oak trembles! Its branches are gone! Its sap is frozen! It bends! It falls! Peace, peace with the white men, is the command of the Great Spirit, and the wish — the last wish — of Passaconnaway."

Wonnalancet now assumed the chieftaincy, and, though he united the wisdom of his father with the virtues of a Christian prince, and uniformly displayed those higher and nobler traits of character worthy of emulation, the doom of his people was even.then foreshadowed; indeed, the "beginning of the end" was apparent. After a sickly existence of a few years, meeting with unexpected cruelty and ill-treatment from those whose superior enlightenment and Christian pretensions he had been taught by the Apostle Eliot to respect and confide in, and being broken and decimated in numbers and spirit, Wonnalancet retired with the remnant of his tribes to the St. Francis, were merged with that people, and there ended his days.

"Wonnalancet" means literally "*breathing pleasantly,*" wonne or wunne (pleasant) and nangshouat (to breathe), and was the name given him after he had come to manhood, when his character was considered as formed, — a name peculiarly appropriate as well as beautiful, as his great wisdom and Christian virtues "breathed pleasantly" on all with whom he came in contact, and over all by whom he was surrounded.

The subjugation of the Indians generally signified not only their reduction to submission, but banishment from the soil, and when this was not the result, some worthless or out-of-the-way portion of their own lands was kindly set off for their occupation and exclusive use.

The densely populated part of the present city of Lowell was thus set off as an Indian reservation. Its boundary on one side was the Merrimack River, and on the other a ditch extending from a point some distance above the Pawtucket Falls to the foot of Hunt's Falls.

When Chelmsford was first incorporated, this reservation was in the eastern portion of the town, and after the departure of the Indians this territory was "squatted" upon by adventurous whites, who were disfranchised and refused representation in the General Court. Whereupon they retorted by politely but firmly refusing to pay any taxes, which brought the matter to a solution by annexation to Chelmsford by act of the Legislature.

It appears that the early settlers and their officials, misled by the course pursued by the Merrimack for thirty or forty miles before reaching the ocean, had adopted the erroneous conclusion that its course was east and west; at least this is the natural inference deduced from documents of the time, official and unofficial. The claim of Massachusetts was to a point three miles north of the most extreme northerly point of the Merrimack River, and, to establish permanently her boundary and extend her jurisdiction, a commission was appointed to perambulate, define, and properly indicate the line to which she laid claim. This commission consisted of Edward Johnson and Simon Willard, who, in the discharge of this duty, determined the outlet of Winnipesaukee Lake to be the point indicated in the claim of Massachusetts. The Endicott Rock, at the Weirs, was marked by the commission as follows: —

EI   SW
WP IOHN
ENDICVT
GOV.

This extraordinary presumption on the part of Massachusetts produced a long and acrimonious controversy between the two States, the absurd claim of Massachusetts being resisted with determination and success.

"Charles I., in the fourth year of his reign, by letters patent, confirmed a grant by the Council of Plymouth to certain persons of a territory thus described, namely: 'All that part of New England, in America, which lies and extends between a great river that is commonly called *Monomack*, alias *Merrimack*,' etc." *

"The Merrimack River was an important boundary in the early times of New England; and it is accordingly frequently mentioned in documents of the times. There seems, however, to have been no uniformity in the spelling. Sometimes, we find several modes in the same document. The modern form has good authority, being emphatically the 'King's English.' The following twenty modes of spelling I have met with in a very limited search; no doubt many others might be added: —

Malamake.	Merremeck.
Maremake.	Merrimac.
Meremack.	Merrimach.
Meremacke.	Merrimack.
Meremak.	Merrimak.
Merimacke.	Merrimeek.
Mermak.	Merrymacke.
Merramack.	Monnomacke.
Merramacke.	Monomack.
Merremacke.	Monumach. †

* Belknap's History of New Hampshire, App.   † James D. Francis, Esq

The period extending from 1790 to 1825 may very properly be called the era of canals. As there were no railroads in those days, this was regarded as the most expeditious and economical mode of transporting heavy merchandise, and canal schemes were as numerous as railroad enterprises at a later day. Canals for manufacturing purposes were constructed around many of the great water-powers, and internal improvements were projected on a magnificent scale; but the Middlesex Canal was the most important of any which were completed, and, with the many others along the course of the Merrimack to facilitate its navigation, was of the greatest consequence to the business interests from Concord to Boston. It may not, however, be generally known that a vast canal scheme, in connection with this, was in contemplation at one time, and surveys were actually made for a line extending westward. Lake Winnipesaukee was to be the base, and, connecting there with the Merrimack River and Middlesex Canal to Boston, it was to extend west by an elaborate system of locking, and ultimately intersect with the New York State canals. But it is probable that the great project of the Manchester and Liverpool Railway, chimerical as it might, and probably did, appear, had the effect to suspend operations on this canal until the result of this new-idea experiment of railroads and locomotive power had been tested. Still the Merrimack, aided by the canals around its falls and the Middlesex Canal, continued to do a heavy transportation business until the railroad was demonstrated to be a success, when, like all other obsolete or old fogy institutions, it gave way to the progress of the age.

In the navigation of the river and auxiliary canals, the family of Tylers bore an important and conspicuous part, and are historically connected with the river and with Lowell. Mr. Nathan Tyler owned nearly all the land from the head of the old Pawtucket, or Navigation Canal, to the Merrimack, and as far down as the mouth of the Concord, the manufacturing companies making their first land purchase of him. His sons, Jonathan, Silas, Nathan, and Ignatius, were born near the foot of Pawtucket Falls, where their father resided and carried on a saw and grist mill, which were swept away by the great freshet. Mr. Joseph Warren was the first superintendent of the Pawtucket Canal, and at his death Mr. Jonathan Tyler was appointed, and had charge of it until it came into the possession of the Land and Water Power Company. Mr. Silas Tyler followed the Merrimack many years as pilot, and was connected with the Middlesex Canal for twenty years, being for seven years captain of packet-boat Gov. Sullivan, and at the opening of the Boston and Lowell Railroad and the consequent abandonment of canal navigation, about 1835, of course his connection with it terminated. Mr. Tyler was of opinion that, with an enlargement of twenty feet in width and three in depth, this canal could have successfully competed with the railroad, as one horse could then haul sixty tons of freight from Lowell to Boston. Mr. Ignatius Tyler has had an uninterrupted connection with the Merrimack and canals from his youth to within a few years, and was for a long time engaged in the lumber trade at Lowell, but for seven or eight years past his operations have been confined to streams whose sources are in the Provinces. When the steamboat enterprise on the Merrimack was started by Messrs. Bradley, Stone, and others, he was captain of the fine little freight and passenger steamer that plied between Lowell and Nashua, and for some years managed an immense carrying trade, via the river and Middlesex Canal, between Concord and all Northern New Hampshire and Boston. He was, like his brother, long connected with the Middlesex Canal, and his employment on the river gave him a familiarity with all the canals around the falls above Lowell. In 1814, the first packet-boat passed through the canal from Boston to Concord, N. H., and in 1819 the first steamboat from Boston reached Concord; and a boat of thirty tons has even gone as far up as the foot of Webster's Falls, in Franklin, the forks of the Merrimack.

A prominent feature of the business of Lowell has been its immense lumber trade. Previous to the construction of these canals, the lumber coming down the river was landed and hauled around the falls to the basin below, which was done by the splendid ox-teams of neighboring farmers; conspicuous among these the old residents do not forget those of Joel and Jonathan Spaulding. Beaver River was also a convenient harbor, and in the season was filled with rafts as far back as the "Navy Yard." Those going via canal to Boston were also hauled by oxen, one pair hauling as many as sixty shooks of rafts. After the Pawtucket Canal was made, this method of shipment around the falls was dispensed with, but the lumber trade was not, and for many years as many as ten million feet of lumber have been brought down from northern New Hampshire by the current of this beautiful river, and find a port of entry here. Besides the lumbering carried on in the early days of Lowell, it was, like all the other great falls along the Merrimack, an important fishing-place. If the savages gathered at this spot for the annual fishing-festival, to secure the food on which they mainly depended, the pale-faces subsequently flocked here with no less eagerness, and the scenes enacted by them would do no discredit to the barbaric orgies of a prior date. The fishing was by law confined to three days each week on the Merrimack and two on the Concord, and officials, known as "fish-wards" were appointed to enforce the law. On the lawful days scores of teams of every description, with drivers in picturesque and ludicrous costumes, came pouring in from all directions, eager to fill their carts for their own use, and for peddling about their neighborhood. The old fish-house, in architecture unlike any other building, with its strange surroundings, its antiquated odor of fish, and an exhilarating effluvia of Old Medford, mingled with the fumes of questionable tobacco, made it compare favorably with Coleridge's description of the city of Cologne, while added to this was the motley group of teams, the busy lumbermen, the excited fishermen dressed for the occasion, cursing, fighting, and plying their vocation, and crowds of spectators enjoying the fun intensely, made up a scene which cannot be described or forgotten by those who witnessed it. At the same time the tavern and the store were doing a thriving business, and all classes of trade received a share of the profits of the fisheries. The leading tavern was kept by Joseph Warren, and after him Jonathan Tyler, who with his wife still lives in Lowell in the enjoyment of plenty and excellent health. This tavern was located on the spot where the American House now stands. The principal store was kept by Phineas Whiting, and was on the site of the elegant residence of Frederick Ayer, Esq. It may not be supposed that fishing was altogether confined to legitimate days; on the contrary, though perhaps for the purpose of keeping the officials busy and zealous in the discharge of their duties, attempts were constantly made at surreptitious fishing and violation of law, which, however, generally involved no more serious consequences than bloody noses, and the engendering of irritation and ill-feeling between the officials and the fishermen. The fish clandestinely obtained seemed to possess a peculiar flavor, or possibly the sport was so attractive to many that legal restraint was impossible. On one occasion, as the disciples of Walton were plying an unlawful business on Long Island (opposite the Lawrence corporation), an obnoxious and officious official from Haverhill (named Vincent), with his posse, pounced upon them, and the scene that ensued may be imagined. Donneybrook was outdone, the official and his party were repeatedly fished out of the river, after unceremonious baptisms by the faithful, and soon as possible beat a precipitate retreat without making any arrests, but with a wholesome lesson in prudence to guide them in the future enforcement of obnoxious laws. It may be thought that these exciting and turbulent scenes were not exactly calculated to cultivate the pillars of a future community; but it must be remembered that the actors were more than half a century behind this enlightened era; that fishing, as practised then, is now unknown; and above all, that every man knew his rights, "and knowing dared maintain" them against all comers, standing more on this point than any particular method or delicacy in the manner of their security. Many of the oldest of the native population of Lowell, not only retain a vivid recollection of these scenes, but were prominent actors in them, and those who would bewail what they choose to term the demoralizing tendency of these disorderly gatherings may see, in the character of the old citizens of Lowell, men of mark and stern virtue, strict integrity, great business capacity, position and influence, a very clear and conclusive refutation of their erroneous conclusions. The family of Tylers, as a sample of the men of Pawtucket anterior to the origin of Lowell, exhibit the quality of humanity indigenous to the soil. For three generations the Tylers of Lowell have been intimately connected with the Merrimack and its canals; in their younger days, in the lively times of Earl Chelmsford, as fishermen, lumbermen, and boatmen; and in maturer years in many high positions of responsibility and trust; and for moral worth, they have had no superiors, and no native or adopted citizens of Lowell have made a fairer record than they, in all the attributes of an honorable and manly character.

The first canal constructed around Pawtucket Falls was to facilitate the navigation of the Merrimack. This was in the year 1792. In 1793 the Middlesex Canal, connecting this river with Boston, was commenced, and its completion rendered the stock of the Navigation Canal valueless; but in the year 1822 it was enlarged for manufacturing purposes. The Merrimack and Hamilton Canals were commenced in the same year. The Western Canal was commenced early in 1828, and the Eastern Canal was built in 1835. The Underground (or Moody Street) Canal — one-fourth of a mile long — was constructed in 1848. The Northern Canal was built in 1846-7, and is one mile in length, one hundred feet wide, and sixteen feet deep. This canal was designed to reduce the current, which was found to be inconveniently strong in the other canals, while at the same time it gave two feet more "head" to some of the mills. The total united length of the Lowell canals is above five miles. The dam at Pawtucket Falls — a substantial stone structure — was built in 1825. The perpetual power at this point of the river is ten thousand horse-power, while the great companies along the lower Merrimack maintain a reserve of more than one hundred miles of surface in the great reservoirs in Northern New Hampshire, which they can reduce, by artificial arrangements, to the depth of several feet, and thus increase the natural and enormous power of the river at pleasure. The perpendicular fall is thirty-four feet; that of Hunt's Falls — two miles below — is ten feet, but is not, at present, considered available for manufacturing purposes.

If aboriginal history and Indian customs give interest to this river, with how much more does civilization and well-directed intelligence invest it! Any one who reads the history of Lowell, especially its *early* history, must feel a glow of pride, if he reads with attention the sketches of the character of its founders as a manufacturing city, which, to be complete, such a work must contain. That the idea of applying the water of the Merrimack River, at this place, to mechanical and manufacturing purposes to the extent and magnitude with which it was applied, and so successfully, too, shows uncommon foresight and judgment in those persons who conceived and carried into execution the great plan. From the first moment of the conception of the great scheme of building up a manufacturing city at these falls no element of character seemed wanting to ensure success. Patrick T. Jackson, Kirk Boott, Nathan Appleton, Francis Cabot Lowell, and their coworkers were emphatically the *founders* of Lowell. They were not merely *builders*, but designers. They had no model either for mills, machinery, or organized system for the practical and profitable application of capital and labor to manufacturing purposes. That men of good judgment, of great mechanical genius and skill, have, from the first, controlled the operations of this vast manufactory, the beautiful and substantial appearance of the mills and dwellings, and the gratifying average rate of dividends bear ample testimony. Perhaps as striking an instance of foresight, or the application of knowledge to a good result, as can be named in the history of Lowell, occurred in the management of the affairs of the "Proprietors of the Locks and Canals," by James B. Francis. Esq., the present chief engineer of this company, a gentleman who has occupied this position for the past *twenty-three years*, and an *employé* of the same company for the past *thirty-four years!* When the *old canal* was enlarged for manufacturing purposes, and the "guard locks" built, the engineer (Mr. Lewis), with P. T. Jackson, Esq., and associates, concluded that the height at which they established them was sufficient to guard against any rise of water that should ever take place, (judging from data which they obtained from old men living in the neighborhood, and from "water marks" which remained of former floods). At the time of building the new (or Northern) Canal (1847–8), it occurred to the agent of the Water-Power Company (Mr. Francis), that as the height of the locks was established to correspond with the water of the river *before* the *river-dam* was built, they were no protection to a portion of the city below them in case a freshet should occur as high as the memorable one of 1785, which was the greatest ever known on the Merrimack since the English settled on its borders. His reasoning was, "what *has* happened, may happen again." Knowing only what the former engineers and agents knew, but excelling them all in the *application* of his knowledge, he caused an embankment to be constructed, six feet in height, composed of masonry and earth, and extending on either side, like a massive brace, to the high land, some twenty rods to the rear of the gateway. Over the canal, through which the boats and rafts must pass, he also caused a huge gate to be suspended by an iron ring or clasp, and gave instructions to the "keeper" of the locks, that if the water should ever rise to a certain point to "send for him, and begin with hammer and chisel to sever the clasp that held the gate." In 1853 occurred the second memorable freshet, which proved the wisdom of his action: the water rose to the point he had indicated; the "keeper" sent for him as directed, and, with hammer and chisel, the gate was dropped into position, thus averting a catastrophe to the city, the consequences of which it is unpleasant to speculate upon. Many of the leading citizens, headed by J. B. French, Esq., desiring to testify in some tangible form their appreciation of his wise forecast, procured a testimonial, suitably inscribed, which they presented to Mr. Francis. In making the necessary excavations for the canal at Lowell the most indubitable evidence is said to have been presented of an important change in the course of the river. Geological indications pointed unmistakably to the fact that at some remote period in the past the waters of the Merrimack were discharged into the ocean at a point not ten miles below Boston; the peculiar stratification, the boulder bed, and the many marks of attrition seemed conclusive proof of this fact.

Nothing can prove more interesting to those who would obtain correct information concerning the city of Lowell than the excellent work of Hon. Nathan Appleton, entitled, "Introduction of the Power Loom, and Origin of Lowell," extracts from which are here presented : —

"At the suggestion of Mr. Charles H. Atherton, of Amherst, N. H., we met him at a fall of the Souhegan River, a few miles from its entrance into the Merrimack, but the power was insufficient for our purpose. This was in September, 1821. In returning we passed the Nashua River, without being aware of the existence of the fall, which has since been made the source of so much power by the Nashua Company. We only saw a small grist-mill standing near the road, in the meadow, with a dam of some six or seven feet. Soon after our return I was at Waltham one day, when I was informed that Mr. Moody had lately been at Salisbury, when Mr. Ezra Worthen, his former partner, said to him, 'I hear Messrs. Jackson & Appleton are looking out for water-power; why don't they buy up the Pawtucket Canal? That would give them the whole power of the Merrimack, with a fall of over thirty feet.' On the strength of this, Mr. Moody had returned to Waltham by that route, and was satisfied of the extent of the power which might be thus obtained." . . . . . "Our first visit to the spot was in the month of November, 1821, and a slight snow covered the ground. The party consisted of Patrick T. Jackson, Kirk Boott, Warren Dutton, Paul Moody, John W. Boott, and myself. We perambulated the grounds, and scanned the capabilities of the place, and the remark was made that some of us might live to see the place contain twenty thousand inhabitants. At that time there were, I think, less than a dozen houses on what now constitutes the city of Lowell, or rather the thickly settled parts of it: that of Nathan Tyler, near the corner of Merrimack and Bridge Streets; that of Josiah Fletcher, near the Boott Mills; the house and store of Phineas Whiting, near Pawtucket Bridge; the house of Mrs. Warren, near what is now Warren Street; the house of Judge Livermore (Edward St. Low), east of Concord River, then called Belvidere, and a few others. Formal articles of association were drawn up, bearing date the 1st of December, 1821. They are recorded in the records of the Merrimack Manufacturing Company." . . . . . "An act of incorporation was granted 5th of February, 1822." . . . . . . "In December, 1822, Messrs. Jackson & Boott were appointed a committee to build a suitable church; and in April, 1824. it was voted that it should be built of stone, not to exceed a cost of nine thousand dollars. This was called St. Anne's Church, in which Mr. Boott, being himself an Episcopalian, was desirous of trying the experiment whether that service could be sustained. It was dedicated by Bishop Griswold." . . . . "The first wheel of the Merrimack Company was set in motion on the 1st of September, 1823. In 1825, five hundred dollars were appropriated for a library. Three additional mills were built. In 1829, one mill was burnt down: in 1853, another. In 1825, Mr. Dutton, going to Europe, Nathan Appleton was appointed president. The first dividend of one hundred dollars per share was made in 1825. They have been regularly continued, with few exceptions, averaging something over twelve per cent. per annum to the present time." . . .

. . . "At the annual meeting at Chelmsford, May 21, 1823, the directors were authorized to petition for an increase of capital to one million two hundred thousand dollars; and, on the 19th of October, 1824, a new subscription of six hundred shares was voted, and a committee appointed to consider the expediency of organizing the Canal Company, by selling them all the land and water power

not required by the Merrimack Manufacturing Company. This committee reported on the 28th of February, 1825, in favor of the measure, which was adopted; and at the same time a subscription was opened, by which twelve hundred shares in the locks and canals were allotted to the holders of that number of shares in the Merrimack Company, share for share. . The locks and canals were thus the owners of all the land and water power in Lowell." . . . .
"The first sale was to the Hamilton Manufacturing Company, in 1825, with a capital of six hundred thousand dollars, afterwards increased to one million two hundred thousand dollars. This company secured the services of Mr. Samuel Batchelder of New Ipswich, who had shown much skill in manufacturing industry. Under his management the power loom was applied to the weaving of twilled and fancy goods with great success. The article of cotton drills, since become so important a commodity in our foreign trade, was first made in this establishment. The Appleton Company and the Lowell Company followed in 1828." . . . . "In 1829, a violent commercial revulsion took place, both in Europe and this country."
. . . . "During this period of depression, Messrs. Amos and Abbott Lawrence were induced, by some tempting reduction in the terms made by the proprietors of the locks and canals, to enter largely into the business; the consequence of which was the establishment of the Suffolk, Tremont, and Lawrence Companies, in 1830. The Boott followed in 1835; the Massachusetts in 1839."
. . . . "In November, 1824, it was voted to petition the Legislature to set off a part of Chelmsford as a separate township. The town of Lowell was incorporated in 1826. It was a matter of some difficulty to fix upon a name for it. I met Mr. Boott one day, when he said to me that the committee of the Legislature were ready to report the bill. It only remained to fill the blank with the name. He said he considered the question narrowed down to two, — Lowell or Derby. I said to him, ' then Lowell by all means,' and Lowell it was. There was a particular propriety in giving it that name, not only from Mr. Francis C. Lowell, who established the system which gave birth to the place, but also from the interest taken by the family. His son, of the name, was for some time treasurer of the Merrimack Company. Mr. John A. Lowell, his nephew, succeeded Mr. Jackson as treasurer of the Waltham Company, and was

for many years treasurer of the Boott and Massachusetts Mills; was largely interested, and a director in several other companies." . . . . "In 1836, the municipal government of Lowell was changed to that of a city." . . . . "The Boston and Lowell Railroad was among the first established in the United States. So early as 1830, a committee was appointed on the subject, and a bonus of one hundred thousand dollars was voted by the Locks and Canals Company, payable on its completion." . . . . . "It was opened for travel in June, 1835, earlier than any other railroad in Massachusetts, for its entire length, and, with the exception of the Camden and Amboy, to Bordentown, in the United States." . . . . "In 1830, Samuel Lawrence, William W. Stone, and others, were incorporated as the Middlesex Company, with a capital of five hundred thousand dollars, — afterwards increased to one million dollars, but subsequently reduced to seven hundred and fifty thousand dollars, — and engaged in the manufacture of broadcloths, cassimeres," etc. . . . . . "The mismanagement of the Middlesex Company's affairs, during many years, was astonishing. The entire capital of the company was lost through the mistakes and irregularities of Samuel Lawrence, William W. Stone, and their associates. In 1858 the company was reorganized, with new managers, and a new subscription of stock. Five hundred shares, of the par value of one hundred dollars each, formed the capital with which the Middlesex Company took their 'new departure' in the voyage of life." *

Mistakes! and irregularities!! This last mild term, it has been not inaptly observed, like charity, covers "a multitude of sins." When an unfortunate individual, under the pressure of the sternest necessity, appropriates or "confiscates" a loaf of bread, or a bundle of fagots, the atrocious! crime is not only characterized in the plainest and severest terms known to the "King's English," but the culprit speedily receives a severe and "wholesome" judicial rebuke; but let some financial Guy Fawkes "blow up" an institution whose capital is represented in millions, leaving the deluded and ruined shareholders buried in the *débris*, and especially if he has plenty of the not exactly definable commodity, professionally or technically termed "character," his nefarious "transactions" are

* "Cowley's Lowell."

gingerly characterized as "irregularities." A poor widow, from dire necessity, purloins a chicken or other food for her famishing children, and justice says, "Six months in the House of Correction" for expiation and reformation; but when some "respectable" speculator, or rather peculator, perpetrates a stupendous fraud or gigantic swindle, regardless alike of duty and honor, bringing ruin to confiding hundreds, justice, if never before, *then* exhibits her proverbial *blindness*.

The following extracts are from "Statistics of Lowell Manufactures, etc., annually compiled from authentic sources by Stone & Huse, January, 1868:—

"MERRIMACK COMPANY. — Capital, two million five hundred dollars; hands employed, two thousand two hundred and fifty; steam, in addition to water power, twenty-two engines. HAMILTON. — Capital, one million two hundred thousand dollars; hands employed, one thousand two hundred and seventy-five; two engines. APPLETON. — Capital, six hundred thousand dollars; hands employed, seven hundred and seventy-two. LOWELL. — Capital, two million dollars; hands employed, one thousand four hundred and fifty; two engines. MIDDLESEX. — Capital, seven hundred and fifty thousand dollars; hands employed, seven hundred and seventy-two. SUFFOLK. — Capital, six hundred thousand dollars; hands employed, nine hundred; one engine. TREMONT. — Capital, six hundred thousand dollars; hands employed, one thousand two hundred and sixty-four. LAWRENCE. — Capital, one million five hundred thousand dollars; hands employed, one thousand seven hundred. BOOTT. — Capital, one million two hundred thousand dollars; hands employed, one thousand three hundred and ten; one engine. MASSACHUSETTS. — Capital, one million eight hundred thousand dollars; hands employed, one thousand seven hundred; one engine. LOWELL BLEACHERY (incorporated, 1832). — Capital, three hundred thousand dollars; hands employed, four hundred; two engines. MACHINE SHOP (incorporated, 1845). — Capital, six hundred thousand dollars; hands employed, six hundred. Total capital invested, thirteen million six hundred and fifty thousand dollars; total number of hands employed, thirteen thousand four hundred and ninety-seven; total steam-power, thirty-two engines.

"The number of churches is twenty-two; school-houses, forty-five; schools, fifty-seven; scholars seven thousand. There are seven banks, (in addition to four savings banks), with an aggregate capital of two million three hundred and fifty thousand dollars. The savings banks have an aggregate deposit of about four million dollars."

To the existence of the splendid water-powers on the Merrimack is largely due the early introduction of the power-loom, together with the extensive and comprehensive system which has made the Merrimack valley and other portions of New England the manufactory of the Western World. It is true that as early as 1787 the power-loom was introduced in Massachusetts, and, subsequently, in a few instances, it was used in Rhode Island and Connecticut; but these cases were few, and the loom so crude and imperfect as to amount to nothing more, either in the way of improved fabrics or increased production, than an experiment; and it was not until several gentlemen, with a determination to make manufacturing an important branch of American industry, took hold of the matter in earnest, and experimented for months, in a room on Broad Street, in Boston, that the power-loom was brought to approach that degree of perfection which it now exhibits. Having by these experiments brought the imperfect power-loom, by improvement, practically to a new invention and a grand success, they submitted it to a practical test at Waltham, where they were incorporated with a capital of four hundred thousand dollars, and proved the new machine to be a perfect success. From the incipient inquiries and investigations into the theory and practice of this branch of business, which was industriously prosecuted not only in this country but in Europe, these individuals manifested a sagacity, talent, and perseverance so peculiar and great as to mark them pre-eminently as the men, probably the only men, competent for the successful prosecution of this new and extensive enterprise.

Starting on a comparatively new and unexplored field, in a business so difficult, complicated, and vast; laying, as they did, the foundation of what may not be improperly called a co-ordinate branch of a nation's industrial interests and prosperity, it may not be too much to say of them that they possessed the highest order of talent and genius, and of either of them, that "his hand the rod of empire might have swayed." Fully realizing the magnificent scheme they had inaugurated in 1814 at Waltham, these sagacious men forthwith turned their attention to the discovery of a motive power commensurate with their requirements, cheap, unfailing, and efficient. The sagacity which had led to successful experiments, inventions, and improvements, and which afterwards resulted equally profitable to themselves and the country, turned their attention to the broad and beautiful Merrimack, in whose great volume and splendid falls they found the motor which possessed the requisite qualification, and around the Falls of Pawtucket the first act of the domestic manufacturing drama of these great actors is seen in the fifty huge factories and the forty thousand people of Lowell, the growth of only a generation. The individuals who were mainly instrumental in the introduction and progress of manufacturing on the Merrimack were Nathan Appleton, Francis Cabot Lowell (in honor of whom the city of Lowell was named), Patrick Tracy Jackson, Kirk Boott, Ezra Worthen, Paul Moody, and others. The two last named, having previously had considerable experience in the business at Amesbury, were practical men and of great aid to the enterprise.

The system which they devised has had able coadjutors, and, though they brought machinery to a marvellous state of perfection, still improvements in minor details have constantly been made; the power-loom, however, except in the substitution of the crank for the cam motion and the increased rate of speed, remains substantially the same. Previous to this time each branch, carding, spinning, weaving, etc., was a separate business, and the transfer of the stock from mill to mill in undergoing the successive processes in the routine of manufacturing, was unnecessarily slow, inconvenient, and expensive, but these individuals perfected an organization of the business which dispensed with these primitive methods by building mills of sufficient size to consummate the whole process under one roof beginning with the raw material and turning out the finished fabric. From this grand beginning at Lowell manufacturing has extended along the Merrimack, above and below, until the valley of the Merrimack, excepting narrow intervening spaces, is one continuous manufacturing community.

"The Wamesits lived at the forks of the Merrimack and Concord Rivers, and upon both sides of the latter river.

"Wamesit is derived from wame (all, or whole), and auke (a place), with the letter *s* thrown in betwixt the two syllables for the sake of the sound. The Indian village at this place undoubtedly received this name from the fact that it was a *large* village, — the place where all the Indians collected together. This was literally true in the spring and autumn, as the Pawtucket Falls near by were one of the most noted fishing-places in New England, where the Indians from far and near gathered together in April and May to catch and dry their year's stock of shad and salmon.

"Wamesit was embraced nearly in the present limits of Lowell, in Middlesex County, Mass."

The Concord River, which unites with the Merrimack at Lowell, is an historical as well as an important stream. Its source is claimed by two towns, and, as the claim of either seems to be well founded, it is but fair to say that it has properly two principal sources. One of the branches of its head-waters rises in a pond in Westboro', and the other in Hopkinton. Uniting, they form an inconsiderable river, which, having its course through a very level country, checks the progress of its current, and thus maintaining full banks gives it the appearance of supplying a much larger quota of water to swell the already plethoric tide of the Merrimack than the fact will warrant.

The Concord is fifty miles in extreme length, and for most of its career it has more the appearance and characteristics of a lake, extensive to be sure, in one direction. Its immobility is unparalleled by any other tributary to the Merrimack; it is dark, sullen, and sluggish, making out into considerable lagoons in places, producing the plants, flowers, fish, and reptiles of the most stagnant and miry ponds, and nothing more. Previous to the invasion and appropriation of these thoroughfares by dams and locks, shad, alewives, and eels, soft fish of the warm-water type, took naturally to this stream; nobler varieties, such as the salmon, avoiding its turbid and uncongenial waters.

Before the advent of the pale-faces this stream was known to the squatter sovereigns along its course by the name of Musketaquid, or Meadow River; the eminent appropriateness of the original name being more distinctly apparent as time rolls on.

In 1635, the town of Concord was settled by English, and the river as well as the town thenceforward took the same name. The year 1775 witnessed scenes here on the very margin of this stream which illustrate the spirit of '76, and gave a direction and impetus to Revolutionary events which disaster, privation, and vicissitudes only accelerated. Called to immediate action, without preparation or efficient implements, these men, on that memorable 20th of April, not so much, perhaps, to oppose tyranny, — for there was not much tyranny to oppose, except in imagination, — but to demonstrate their faith in self-government, and determination, come weal or woe, to bring it to a practical test, — those men, all untrained and inefficiently armed as they were, threw themselves into the "imminent deadly breach" at the old north bridge, and by their valor, by indomitable determination and pluck, forced the serried ranks of "perfidious" Albion's veterans to waver, yield, and fly.

The abutments of the battle-ground bridge still remain almost the sole surviving relics of this auspicious inauguration conflict.

The falls on the Concord River at Billerica have long been used for mechanical and manufacturing purposes, and were an important element in the prosperity of that section more than a century before ground was broken for manufacturing purposes at Lowell.

As early as 1708 the town authorized the construction of a dam at this place, which was undoubtedly the first ever built on this stream. This dam was constructed for the indispensable saw and grain mills of the early time, and was likewise, at various periods, used for other purposes; there being at one time a comb factory and some other mechanical works which obtained their motive power from this fall.

In 1740, a mill was erected, of considerable consequence at that early period, filled with machinery for finishing home-made cloth, called the "Fulling Mill," and was operated for a number of years, when it was purchased by Francis Faulkner, Esq. It was operated several years by Messrs. Faulkner & Son, and has continued uninterruptedly under the control of the family to the present time, having been since 1810 under the immediate management of J. R. Faulkner, Esq., the present proprietor.

In 1836, the old mill was destroyed by fire, but was rebuilt immediately, with four sets of machinery, and in 1865 the mill was enlarged, and three sets additional, making seven sets altogether, were put in operation. The average product of this mill is twenty-seven hundred yards of flannel daily.

In 1835, there was an old shop at this place used for the preparation of dye-stuffs, which was, in 1839, leased by the Messrs. Talbot who continued the preparation of dye-woods, drugs, etc., on a small scale at first, but with judicious management it steadily increased, and grew to be an extensive and important branch of business.

In 1840, a carpet manufactory was established here by Messrs. Lang & Lannagan, which being destroyed by fire was not rebuilt. In 1851, Messrs. C. P. Talbot & Co. purchased a franchise of the *Middlesex Canal Co.*, and thus secured a priority of right to the mill privilege, which is rated at two hundred horse-power, and, in 1857, erected a fine mill for the manufacture of flannels, which they have since continued in successful operation, the average daily product being three thousand yards. As has already been observed, the Indian name of this stream, Musketaquid or Meadow ·River, was peculiarly appropriate, and eminently descriptive of its characteristics; the country through which it passes being so remarkably level that its waters are sullen, sluggish, and slimy, making out through the numerous depressions of the soil in extensive marshes and lagoons, foul with rank water-grasses and filthy reptiles, the haunt and fishing-ground of the majestic bittern, which, though a native of Europe, is found plentiful in all the fresh-water marshes of New England; the black tortoise, the spotted fresh-water terrapin, the great water-adder, and the hordes of disgusting water-bred reptiles common to stagnant pools; its redeeming features being the fine farms and broad green fertile intervals which border it, and the "milky way" starred with the great white water-lily of midsummer, which in unequalled beauty and fragrance floats gracefully upon its tranquil surface.

The character of the Concord is entirely different from any other tributary of the Merrimack; indeed, a gentleman who is something of a naturalist, and a careful observer, previous to the building of the dam at Lawrence, and when the tribes of the sea came up the Merrimack, noticed the difference between the shad, alewives, etc., which passed into the tepid, stagnant waters of the Concord, and those of the same variety, which, proceeding up the Merrimack, sought its livelier affluents, — the former being darker in color, softer, poorer in flavor, and less palatable than the latter. This being the natural condition of the river, it is not a matter of surprise that the landed proprietors along its course, who had come many years subsequent to the construction of the Billerica Dam, should naturally attribute the stagnation and flowage to that, instead of natural causes. Such was the fact, and the proprietors of interval lands in Sudbury, Concord, and other towns on the river, honestly believed that the dam was a positive and serious detriment to their lands along the margin of the stream; and as early as 1810 a suit was brought against the proprietors of the mill privilege for damage, and to restrain them from the flowage of these lands.

This suit was decided in favor of the mill owners; but this decision not being satisfactory, another suit was brought two years later for the same purpose, which was also decided adversely to the parties claiming to be aggrieved by the maintenance of the dam at Billerica. This put a quietus to the water controversy for many years, and peace reigned on the Concord. In 1828 the old wooden dam was replaced by a substantial stone dam, and the enterprising proprietors kept the wheels of their machinery steadily in motion. The feeling, however, which had been entertained for many years that the backset of this dam was a nuisance and a serious damage, was not obliterated by adverse judicial decisions, nor even abated; the belief prevailed that it was a grievance which could not, and ought not to be borne. Accordingly in 1862, the question was brought before the Massachusetts Legislature, with a fixed purpose, on both sides, of giving the whole subject a thorough investigation, with a view to a final adjustment. The land owners argued, that, if the mill owners had bought the Middlesex Canal, it did not follow that they had a right to flow the lands; in fact, they had no right, and could acquire no right to damage their property, at least, without ample and satisfactory compensation. On the other hand, it was proved that in 1798 an addition was made to the original charter of the Middlesex Canal Company, granting them a right to use the water of the Concord River for manufacturing or mechanical purposes, and when the franchise of the Middlesex Canal was purchased by the Messrs. Talbot & Co., the *water-power right* was included in the purchase. It was contended that the dam did not cause the overflow of the lands in question. The Legislature appointed commissioners, engineers, and other gentlemen fully competent, to settle the question on its merits. The commissioners, assuming the entire control of the water of the river at and above the Billerica Dam for a period of thirty days, made many thousand measurements of the water in or under every conceivable varying condition, and reaching far up the stream, opening the sluices and reducing it to its natural current; in short, making as thorough an investigation as scientific engineering skill could possibly make, with a view to elucidate the facts. The board of commissioners, having given the whole subject a most searching and unbiased examination, reported in effect, as is understood, that the dam at Billerica was not the cause of so much of the lands bordering on the river being water-logged and ruined; that the cause of this condition of the meadows was not owing to the maintenance of the dam: that it was entirely independent of any artificial obstruction, but was precisely the natural condition of the river, and the lands adjacent, and nothing else; and the Legislature confirmed this report by a vote to that effect. In the legislative decision of this celebrated case, so important and interesting to all proprietors of water-power in New England, after a full, patient, and careful investigation, all parties concerned, it is hoped and believed, finally and fully acquiesced; the land owners having stood firm, with valor and determination, for their rights until they were defined and established by a competent tribunal; while the Messrs. Talbot and Faulkner defended and maintained the rights of the proprietors, not only of this water-power, but all other water-powers, at an enormous expense, and with a zeal, energy, and determination which proved that they fully realized the importance of the issue, and comprehended the disastrous consequences not only to themselves, but to the entire manufacturing interests of New England, involved in an unfavorable result of this great contest.

The torpid character, which has already been described, is maintained by the Concord, with the exception of the falls at Billerica, until it arrives in the vicinity of that part of Lowell known as Whipple's Mills. If it has been a marvel of idleness and stupidity for the first fifty miles of its career, it makes an earnest effort to compensate, as far as possible, by its great activity throughout the remainder of its course of about two miles, to the Merrimack. From the point indicated it is little else than a constant succession of rapids and falls, which are generally improved by men of sagacity, enterprise, and activity, and made to contribute largely to the wealth, industry, importance, and growth of the young and flourishing city of Lowell.

As early as 1801, Mr. Moses Hale built and put in operation a carding-mill a short distance from the Concord, on *River Meadow*, or, as it has long been called, Hale's Brook, which falls into the Concord at the foot of Whipple's Falls. In 1818, Mr. Hale established works for the manufacture of gunpowder at these falls, and took into his employ an athletic young man, whose name was Oliver M. Whipple, who proved to be as robust in intellect, determination, and self-reliance as he was physically, and the following year he was made a partner; and, managing the business with great ability and success for many years, he at length became sole proprietor, and continued this hazardous business with unabated success until the year 1855, when, in consequence of ill-health, he disposed of stock, fixtures, and machinery, which were removed to the State of Maine and put in operation by another company. Mr. Whipple early saw the importance and value of these falls for manufacturing and mechanical purposes, and at once adopted measures to secure these advantages. To accomplish this purpose, he constructed a canal from the head of the falls, nearly parallel with, and a few rods distant from, the left bank of the river to the foot of the falls, thence taking a westerly course, discharging its waters, after being used into *River Meadow Brook*. The total perpendicular fall at this place is twenty-five feet, affording excellent mill sites, which are occupied by Faulkner's Mill, flannel; Chase's Mill, fancy woollen; Charles A. Stott's, flannel; American Boot Company; Belvidere Woollen Company, flannels; Shuttle Factory; American Bunting Company; Naylor's Carpet Company; Grist Mill and Worsted Mill. This canal is claimed to be the first of the kind ever constructed for such uses in this country; and, so chimerical did this project appear to the partners of Mr. Whipple, that the eminent engineer, Loami Baldwin, was called here to give his opinion in regard to it. He at a glance saw its importance and valuable arrangement to apply the water-power to proper uses. Of course, having been examined and decided favorably by Mr. Whipple's comprehensive mind, no other result could have been reasonably anticipated; thus the only effect of this investigation was to confirm the wisdom of Mr. Whipple's conclusions by testing them with eminent engineering skill. In September, 1821, operations were commenced for the construction of this canal, which was hastened to its completion; and it is not improbable, in fact, it is known,

that this small canal had a favorable influence on the men who, the
following year, examined the Pawtucket Falls with a view to establishing the immense business which pre-eminently entitles Lowell to
the distinctive name it bears, — the City of Spindles. Thus, it
will be seen that Oliver M. Whipple is entitled, perhaps equally with
the other distinguished gentlemen whose names have already been
given, to the credit of founding, as he certainly is, in his individual
capacity, to the building up of a famous and flourishing city; and no
citizen of Lowell, or elsewhere, can fail to appreciate the conspicuous part acted by him in forming the worthy character and establishing many of those successful enterprises which redound to the
prosperity and fame of the city of his adoption. While on the floodtide of prosperity in the manufacture of gunpowder, Mr. Whipple
found time to pay some attention to such improvements as would in
the future be of advantage not only to himself, but to individuals
and the community in which he lived. With these aims constantly
in view, and actuated by such motives, he turned his attention
steadily to securing such of the lands bordering on the river as were
necessary for securing the power and independent control of the mill-sites which it furnishes. In this manner, and by degrees, he obtained possession of much of the land, but it was not until after the
year 1834 that he secured the lands on the east side of the river, and
with them the entire control of this water-power. Mr. Whipple has already disposed of the right to six hundred horse-power to different individuals, while the capacity of the stream is not yet nearly exhausted.
Mr. Whipple deserves more than a passing notice, as he has always
been a very enterprising man, ready for any position in public life
to which the people of Lowell have called him, and efficient in every
position; public-spirited, generous, and coming from a stock that
bears the stamp of honesty, integrity, and uncontrollable activity,
he is a worthy representative of the name he bears, and of the race
to which he belongs. He is now living in the enjoyment of good
health and a vigorous intellect, and takes great pleasure in reviewing
the stirring events of the last fifty years of his life, — events in which
himself has been one of the most prominent actors, — which he can
review with gratification and the utmost self-satisfaction, as everything of an important character in which he has been engaged bears

the flattering testimony of a wise forecast, an unusual prudence, and skilful management in its beneficent and successful result.

On the 6th of April, 1854, the sons of Vermont, residing in Lowell, held a pleasant reunion, and of this meeting Oliver M. Whipple, Esq., was president. By referring to the columns of the "Lowell Courier," in which the proceedings of the meeting were published, it is seen that the chairman, Mr. Whipple, was born in Weathersfield, Windsor County, Vt. From remarks made on that occasion, it appears that, having reached his majority, with such education as three months of annual common schooling afforded, a liberal supply of good common sense, and fifteen dollars in cash as his stock in trade, Mr. Whipple, in 1815, set off on foot, as light in heart as in pocket, to seek his fortune, shaping his course for Boston, which city he reached after a diligent and tedious journey of four days. Arriving in Boston he was soon employed by parties, and proceeded to Southboro', where he remained three years, and by industry, economy, and frugality he was enabled, in 1818, to go to Lowell with the snug little sum of six hundred dollars in his pocket. At the time he commenced operations on the Concord River the place was known as East Chelmsford, and contained but a few houses; he has, therefore, a historical connection with the city of Lowell from its origin to the present time, and is certainly, as a business man and a citizen, one of its most important features. From this humble beginning, in 1818, with a total cash capital of six hundred dollars, some idea may be formed of his industry and enterprise, when it is known that, in 1856, the assessors' book showed that his annual tax was fifteen hundred dollars; and it must be a satisfaction to him to reflect that he has *earned* his ample fortune, — which is more than some other prominent Lowellians are popularly credited with.

Although Mr. Whipple, by his ability and untiring energy, has deserved and enjoyed a large measure of material prosperity, still no efforts of his could ward off those chastening afflictions sent by a mysterious hand, and which are the lot of a common humanity. About 1852, under the long-continued pressure of his arduous duties, his strong constitution gave way, and for several years his health was so precarious that no one, except himself, regarded the prospect of his recovery as hopeful. He, however, relying upon a naturally robust *physique*, and regular, abstemious habits, seems to have been the only one who never doubted a favorable final result. Unfortunately, during this trying period there came a general stagnation in his line of business, which would have involved men of smaller calibre than Mr. Whipple in utter and irretrievable ruin. But that remarkable sagacity which had been his passport to success remained in sickness and adversity one of the chief elements of his character. He saw that his best move was to render as much as possible of his real property available, and for this purpose secured the valuable services of E. B. Patch, Esq., for thirty-five years a resident of Lowell, and who had been largely engaged in the commission and real estate business, and who was known to Mr. Whipple to be a sagacious and prudent manager. Mr. Patch, as agent for Mr. Whipple, entered upon the active discharge of his duties, when the latter realized sufficient from his advantageous transactions to set him financially on his feet again; but, ill health continuing, Mr. Whipple found it advisable to dispose of his powder works, and afterwards the balance of his unoccupied water-power and other unproductive real estate.

In 1861, Mr. Patch secured a bond (for a deed) of all the unsold water-power on the Concord River, and the lands below the Cemetery Bridge, on both sides of that stream, to the confluence of Hale's Brook, — including the grove on both sides of Lawrence Street and the land extending west to Central and Gorham Streets, — nearly fifty acres in all, which he laid out in building and factory lots, by which means he disposed of them and the unused water-powers, the former for dwellings, the latter to A. H. Chase, the Lowell Bleachery Company, N. R. Wood, S. W. Faulkner, Hosford E. Chase & Co., for manufacturing purposes. In March, 1863, Mr. Patch took a deed of the property remaining unsold, and for the next two years devoted his time and his energies to the development of its resources. During this period wonderful improvements were made on this neglected territory; the canal was much enlarged, its capacity being nearly doubled, its banks and gates were repaired, twelve additional wheels were set in motion, buildings were altered and repaired and new ones erected, machinery was overhauled and new and improved patterns substituted, streets were laid out, bridges built, and the facilities for business generally were so increased that the rents advanced from about three thousand dollars to nearly ten thousand dollars per year. He also gave the city land and material for widening and repairing Lawrence Street from the canal to the Cemetery Bridge, making it a pleasant drive, and as fine a piece of road as any in Lowell.

In 1865, Mr. Patch disposed of the balance of his property to General B. F. Butler, who organized a company at once, called the Wamesit Water-Power Company. This brief synopsis of his transactions shows what a man of energy, ability, and integrity, like Mr. Patch, can accomplish, and the measure of his merits may be better estimated when we consider how at that time the nation was in the midst of a gigantic struggle; a terrible financial revulsion was impending, and a general prostration of business prevailed, not only in Lowell, but over all the land. Overcoming all these serious obstacles, which would have appalled men of less activity and determination, he pushed steadily forward until his purpose was splendidly accomplished. Mr. Whipple, almost on the verge of the grave, and totally incapable of managing his affairs, yet exhibited his superior judgment in the selection of Mr. Patch as manager, and he, in return, demonstrated the wisdom of the selection. Comparatively valueless property was brought to a productive condition, many thousands of dollars were put in deserving pockets, and consequently added to the taxable property of Lowell; and as a proof of the strict and rare integrity of the parties interested it may be said truly that no two gentlemen enjoy each other's respect and confidence more fully than Mr. Whipple and Mr. Patch, and nothing could be more fortunate or gratifying to both than the result of their business connection. Mr. Whipple became fully restored to health, and had the good fortune to find the crisis in his financial affairs averted, and his ample fortune unimpaired; while Mr. Patch had the great satisfaction of realizing that he had not only averted a sad calamity to an old and valued friend, but at the same time, by his talent and judgment, he had built up a business territory which is a credit and a marked feature of this enterprising city, and at the same time had secured for his toil and enterprise an ample, material compensation.

There are few streams of its size, so romantically historical in its records, traditions, and surroundings, and in its application to the service of man and the promotion of useful arts, as the Concord. For fifty years past the true character of this remarkable stream was so little understood even by the land-owners along its borders that an almost continuous and very acrimonious legal controversy was maintained, which resulted in establishing the fact, by an able Board of Legislative Commissioners, that the river was a very different thing from what they had, all their lives, supposed it to be. An inability to understand its true character had always prevailed. It had been an aggravating and expensive problem to some, and an insoluble mystery to others. The first blood of the Revolution, the blood of the intrepid and invincible though untrained yeomanry, mingled with its turbid waters at the Old North Bridge, and long years before it had been the haunt of the wary and stealthy barbarian, who, swooping down upon the exposed and defenceless settlers, enacted those atrocities which marked the advancing borders of civilization in New England, and makes the history of that epoch a yet existing terror. It was then called the Musketaquid or Meadow River, and it is the meadow river still, — a strong proof that the appropriateness of Indian designations need not be questioned, much less changed. If, in some sense, a river is the type of human life, this particular stream may be cited as symbolizing the actual career of many individuals known to those who may give the comparison a little reflection. How many there are who start off on the journey of life like this stream, — useless, idle, and aimless, instead of becoming a wheel, a lever, an axle, a *something* in that complicated machine called society ! The topography of the country is such, and the aspect of the stream so peculiar, as to warrant the supposition that it had repudiated natural laws, ignored the attraction of gravitation, and had taken its course over a gentle acclivity, which has the effect to get itself repudiated in turn by those same laws, as it leaves its bank through every depression, and ruins much of the adjacent soil by the creation of swamps, marshes, and lagoons. Thus it is with individual idleness, disfiguring the course of life with waste places, while the sedges, rank water-weeds, and ugly, filthy reptiles represent the vices, little and great, the fungi bred by indolence, — a parasitic growth.

On reaching Billerica, as if disgusted with its previous lack of energy, the stream makes a daring though spasmodic attempt at reformation, but after this tolerably successful effort, is apparently satisfied or exhausted and relapses into its former state of stupidity, which it maintains until it arrives at maturity, and at Lowell, where, like the man who awakes to a realizing sense of his duties, obligations, and responsibilities at the eleventh hour, throws off the lethargy that has held it so long in chains, and, dashing over nearly two miles of picturesque and powerful falls, seems to seek, and with entire success, to compensate for its former vagrant life, and finally throws itself with alacrity into the Merrimack, leaving no space between the termination of its beneficent labors and its final doom; typifying not only the necessity but the grandeur of a reformation which, by earnest, vigorous works, testifies to an ultimate appreciation of the duties, objects, and obligations imposed by the very fact of the creation of capacity and inherent power. The deep tinge of romance surrounding this stream in its native condition has not faded or diminished, but is, rather, intensified by the peculiarities of the men and the circumstances connected with the inauguration and prosecution of improvements around its splendid waterfalls.

Nearly one hundred years ago a man named Ezekiel Hale, who resided at West Newbury, possessing a great and almost purely mechanical genius, left his home in search of a water-power available for the development of a manufacturing scheme, — a branch of industry then in its infancy in this country, which his vigorous intellect had long and anxiously studied; an enterprise crude and meagre, to be sure, compared with the gigantic achievements of the present day, yet, for that early time, a grand conception. Having, on his way, examined the capacity of Little River, in Haverhill, and concluding to look farther, he finally set down on the lower falls of Beaver River, in Dracut, where he laid the foundation for the extensive Merrimack Woollen Mills, which now occupy the site where Mr. Hale begun. His sons, imbued with the same spirit as the father, now cast about for a new and independent beginning, each for himself. Ezekiel returned to Little River and established himself where E. J. M. Hale, Esq., has continued the business for many years until recently, — being superseded by Mr. Stevens, a connection of the family, when Mr. Hale engaged in the manufacturing business, more extensively, at Groveland. Moses, another son of Ezekiel, senior, prospected for himself. In his peregrinations he crossed the Merrimack and found himself in the neighborhood of Whipple's Falls in the Concord River, and exclaiming "Eureka!" with the enthusiasm of a devotee, he pondered on this prodigal waste of power until the sun descended behind a crimson drapery of clouds, the gorgeous, purple twilight deepened into gloom and "left the world to darkness and to" him. Thoroughly mechanical in his intellectual organization, he could not fail to realize that here was an agency with the capacity to expand and mature his preconceived desire to found a mechanical and manufacturing community himself. Composed of speculative as well as operative elements, his mind here recognized a harmonious blending of those qualities for which he yearned, and was quieted. This was his place. At the head of these falls stood a beautiful perennial grove, its stately and venerable columns supporting a dense canopy of foliage, and whose gloomy, silent avenues, secluded dells, and mysterious grottos might be peopled with the *genii* of this romantic region. Who shall say, in his dream of those grand results he scarcely dared anticipate, — the working of the irrepressible principle of progression, — but Vulcan stood upon the border of this terrestrial paradise of shadows with bare and brawny arm and ponderous hammer, ready to summon his *corps d'industrie* to the furnace, and to forge the manacles which should securely bind this hydraulic giant, and cast him, prone and pliant, at the feet of this Divinity of Mechanism, whose name was Moses Hale? Who can say but Venus, in obedience to the will of her lord, assembled her fair daughters and explained to them the new path and the new duties about to open to them; how nimble and cunning fingers must manipulate the machinery, and in the production of fabrics only yet foreseen, and, with an admonition to be faithful, gave them the maternal blessing? Perhaps this very circumstance was the foundation for that display of these same daughters, unequalled in magnificence, which, many years later, in 1833, welcomed the Hero of the Hermitage, on his memorable visit to the spot which had already come to be known as "the City of Spindles." With his characteristic gallantry, Gen. Jackson visited Lowell to pay his respects to these beautiful operatives, the fair daughters of New England, who had left their quiet country homes and come here to astonish the land with the amazing growth of this new city and new branch of industry and the large production of cotton fabrics reached in so short a period of time. On the 26th of June, 1833, Andrew Jackson, President of the United States, accompanied by the Vice-

President and several members of his cabinet, visited Lowell. His desire to journey hither was quite natural, he, having some eighteen years before felt the protecting power of cotton breastworks, doubtless had a great curiosity to see the material of his famous fortifications transformed into equally famous Merrimack prints and sheetings by these intelligent girls. The distinguished party was received and addressed by the town authorities, and the president responded. Triumphal arches of evergreens, banners, and flowers had been erected, and the escort was composed of "the selectmen, committee of arrangements, Kirk Boott, chairman, a regiment of militia, a cavalcade of two hundred citizens, six hundred school children, and two thousand five hundred factory girls." As each girl was dressed in holiday attire of unrelieved white, the vast procession of Lowell's beauty and worth made a deep and enduring impression on all who had the good fortune to witness it; and it is not a wonder that the Hero of New Orleans, who had often met and overthrown a hostile foe, should evince a willingness to surrender for the first time, without any further hostile demonstration from this all-conquering array than presenting arms. After the public ceremonies, a number of the girls, still in gala dress, repaired to the Merrimack Mills and initiated the president into the art and mystery of making cotton cloth, — the process itself demonstrating the immeasurable difference between the application of this staple in his hands to the necessities of barbaric war, and in theirs applied to pleasant and profitable pursuits of peace. M. Chevalier, an intelligent French gentleman, was of the president's party, and the splendid pageant made a deep and lasting impression on his mind; and, as a high official of the French government, he addressed one of the Massachusetts senators with a view to securing the attendance of a group of those operatives, with the necessary machinery, at the recent Paris Exposition, where the superb New England factory girl at her vocation, and the product of her skill and industry, could be seen together by a world's admiring representatives. Thus early in the history of manufacturing in this country, and of Lowell, had these fair daughters acquired a transatlantic fame, and no one doubted, however visionary and ideal the scheme, and uncertain and doubtful the results of its founding, it was now a well-established fact.

Mr. Hale, without a clear or well-outlined idea, perhaps, of what was to be accomplished, though with a settled faith that his purpose, still only faintly outlined in his own mind, would mature, and the evidence of his creative genius become familiar to men by plain, outward, and visible signs, while the advantages should be great, and benefit the race located permanently here. Scarcely more than a myth himself, and his project a kind of phantasy, he was yet, when securely enthroned on this then solitary spot, the embodiment of the Spirit of Progress, — the wizard who should wield the enchanter's wand, and the waste places were to be occupied by a people characterized as enterprising, skilful, industrious, and thrifty. It was not long before his very eccentricity assumed a tangible and practical shape, and, in 1795, he set a carding-mill in operation. Soon after, expressing an intention to extend and increase his business in the same direction, he was taken aside by a friend and gravely advised to desist, as the capacity of the works already in operation was sufficient to amply accommodate all Middlesex County, and such a reckless investment would not only exhibit a stupid oversight, and lack of judgment and sound discretion, but would involve him in inextricable financial ruin. Nevertheless, he continued to increase his business, and of course his income. Mr. Hale dug a canal in Chelmsford, which was, considered as an individual enterprise, an operation of considerable magnitude, and displayed his interest and belief in internal improvements. It is said, also, that he conceived and suggested the idea of constructing another canal, which was acted upon some years after, the canal being now a "power in the land." Shops and mills began to multiply about the Wamesit Falls, dwellings followed the house of religious worship, and the courthouse and the cemetery have appeared, — emblems of a high civilization. As Lowell rapidly increased, by the improvement of its splendid water-power, in population and productiveness, it began to attract the attention of economists at home and abroad, and, in 1834, an agent was sent to the United States by the "Citizen King," to inspect the public works of this country. He visited Lowell, its shops and factories, and, of the impressions there received, M. Chevalier wrote to the Paris "Journal des Debats" as follows : —

"Unlike the cities of Europe, which were built by some demigod, son of Jupiter, or by some hero of the siege of Troy, or by an inspiration of the genius of a Cæsar or an Alexander, or by the assistance of some holy monk, attracting crowds by his miracles, or by the caprice of some great king, like Louis XIV., or Frederick, or by an edict of Peter the Great, it (Lowell) is neither a pious foundation, a refuge of the persecuted, nor a military post. It is *a speculation of the merchants of Boston.* The same spirit of enterprise which the last year suggested to them to send a cargo of ice to Calcutta, that Lord William Bentwick and the nabobs of the India Company might drink their wine cool, has led them to build a city, wholly at their expense, with all the edifices required by an advanced civilization, for the purpose of manufacturing cotton cloths and printed calicoes. They have succeeded, as they usually do, in their speculations. The inhabitants possess in the highest degree a genius for mechanics. They are patient, skilful, full of invention; they must increase in manufactures. It is, in fact, already done, and Lowell is a little Manchester."

In the immediate vicinity of the head of the falls, on the Concord, is situated the Lowell cemetery, — the beautiful City of the Dead, — extensive in territorial area, romantic in its surroundings of river, hill, forest, and field, and also rich in ornamental trees, shrubbery, and flowers. Its avenues, aisles, and graves are covered with the shadows of luxuriant foliage, and fringed with shrubs, plants, and flowers; wild birds carol sweetest melody among the trees, in careless, joyous, unrestrained freedom, "so clear, that men and angels might delight to hear." The grounds are tastefully laid out and adorned with luxuriant shrubbery, while the countless cultivated annuals which fleck the surface, and the sweet, modest little wild blossoms mingle their perfume and freight the air with delicious fragrance. The hand of affection has inscribed the record of the deeds and worth of the departed on tablets of stone and monuments of enduring marble; a deep stillness, an all-pervading quiet — save the wind rustling the foliage, and the solemn sighing of sombre perennials, a perpetual requiem — rests upon the place; a silence such as might be gladly sought to lull one to eternal repose, enchanting prelude to that final sleep that knows no dream, and knows no earthly waking; a symbol of the solitude, though now adorned and beautified, which covered all this realm around the fine falls of the Concord, when it was invaded by the all-subduing and all-conquering

Divinity of Mechanism, in the time long agone. And though Mr. Hale has departed, and many others have gone, and the populous village of silence, of shadows, and of human ashes, have come of their going, still the places they vacated have been filled again by active, earnest, living souls, and his mantle has fallen upon the shoulders of, and is gracefully worn by, his son and successor, B. S. Hale. In him the character and genius, as well as the name, are still preserved. The spirit which was once so potent here is still believed to hover around and furnish inspiration. Mr. Hale is largely engaged in the manufacture of laid cord, — a most convenient means by which to bind the aims and ends, the objects and interests of mankind together. There is the clock cord, which, moving the finger on the dial, tells off the knell of time; the cord "to hold as 'twere the mirror up to nature;" the picture-cord, to hold the shadow of the beloved and departed substance up to view; the chalk-line, marking the right line of duty which every man should hew square up to; the fish-line, suggesting how, with gilded lure, the evil one is said to angle for the souls — and bodies — of those he has already set his mark upon. Formerly the method of laying cord was by hand, and many inventions have been devised to render the manufacture more rapid, but the cord was laid irregular and uneven, and no permanent success attended the invention of machinery for this purpose until Mr. Hale perfected the machines now in operation at his mill, by which the cord is laid more uniform than even by hand. Thus has the Concord, when its course is nearly run, accomplished its redemption, and, as a man's life is often likened to a river, it may be well for those who can, to redeem, as the Concord has, a vagabond character, even at the eleventh hour, and thus maintain the parallel.

The Wameset Water-Power Company having purchased (February 7) the balance of the unimproved power at Whipple's Falls of E. B. Patch, Esq., the company was organized May 20, 1865, as follows: Hon. Tappan Wentworth, President; Gen. B. F. Butler, Treasurer; D. C. G. Field, Agent; capital, one hundred and fifty thousand dollars. The company set about utilizing the power with great activity, and the genius of a master mind was seen on every hand. A substantial brick mill has been erected (four stories), occupied by the United States Bunting Company and Wamesit Worsted Company; a stone mill (two stories), by the United States Cartridge Company, fixed ammunition; a stone mill (four stories), Worsted Company; a wooden mill (two stories), carpets and dry house; also, a large boarding-house. These mills are driven by two turbine wheels of two hundred and thirty-one horse-power, with a reserve engine of one hundred and fifty horse-power, to be used in the event of low water. Two boilers, five feet in diameter and sixteen feet in length, each with eighty tubes, supply the steam. A branch of the Boston and Lowell Railroad affords easy transportation directly to and from the mills. It has been found that much of the available power is lost by the incapacity of the canal, and a saving enlargement is projected the coming season, so that mechanics of all kinds can here find permanent room and power to prosecute their business, which will ultimately build up around these falls a large and important manufacturing village.

The political and military history of Gen. B. F. Butler is well known to the world, but of his extensive engagement in the pleasant, peaceful pursuits which adorn, dignify, and enrich a community, little, perhaps too little, is known to the general public. Upon the reorganization of the defunct Middlesex Company, Gen. Butler became a large purchaser of the new stock, and, with his characteristic energy, devoted much time to the practical details of manufacturing. When the Whipple Falls property was put in the market, Gen. Butler, seeing its real value and importance for manufacturing and mechanical purposes, lost no time in securing it on his own account, and forthwith adopted measures for the founding of those enterprises which have already added largely to the business and prosperity of Lowell, and places Gen. Butler among the leading and successful manufacturers of New England, while it presents him as a promoter and patron of those arts which, while they give comfort, industry, and thrift to a community, afford additional stability and perpetuity to its institutions. Men of great means are not always men of great deeds, and it is certainly refreshing to contemplate a public-spirited and liberal man whose means are equal to his disposition, and who, in his investments, has the sagacity to discriminate in favor of such substantial interests as will surely tend to the present and prospective advantage of the place and the people. The difference between a comprehensive and a superficial mind is as marked as between the generous and sordid, and is patent to all. While the one is satisfied with a resort to temporary shifts and expedients, the other builds on a solid foundation, which is not only a warrant for current profit, but a bond for future premiums and dividends on such judicious investments. The great hydraulic power of this fall was well known, but the expenditures necessary to its productiveness, for mills, machinery, appliances, and stock, before the dollar could return to their pockets was too appalling for some gentlemen, who decided to cling to their bonds and greenbacks, and it was not until Gen. Butler, on a brief visit from the "tented field," learned it was in the market, when he purchased it, and initiated measures which must not only advance the growth and character of the city, but "put money in his purse" for many years to come. He being the principal owner exercises a controlling influence over the operations of the company, and, having selected a gentleman of peculiar fitness for acting manager, in the person of Mr. Field, this company is now on the road to prosperity, and bids fair to become one of the most prosperous and noted companies along the Merrimack. In the brief history of this company, here given, Gen. Butler is seen as a practical manufacturer, and, however opinions may differ on other points, there will be a unanimity in the recognition of the great credit which attaches to him for his energy, comprehensive views, for his wise and liberal use of an ample fortune, for his active public spirit, exhibited in building up the industrial interests of his adopted city, and his generous liberality in the promotion of her general welfare.

The United States Bunting Company, one of the enterprises originated since the Butler purchase, is deserving of special notice. Previous to the establishment of its works the manufacture of Amercan flags from an American fabric did not exist in the United States, all bunting having been previously imported. No American ship or American soldier had ever fought a battle under a flag of American manufacture. Many attempts had been made to produce it, but without success. At the solicitation of several of the government departments, and encouraged by the act of Congress of 1865, empowering heads of the several bureaus to purchase bunting of American manufacture at the same price as the imported article, several gentlemen in Lowell associated for its manufacture. After many experiments they succeeded in making an article of American bunting, which, after being put to the severest test, was certified by the officers of the Navigation Bureau of the Navy Department to be superior to the English manufacture in texture, material, color, and durability. Since then the bunting of this company has been used by all the government departments, to the exclusion of foreign manufactures. Stimulated by their success, this company has erected large and permanent mills, and supplied themselves with requisite machinery of sufficient capacity to supply the wants of the American people.

During the progress of this manufacture this company has made many improvements in the fabric and the machinery for its manufacture. One of the most prominent is the manufacture of buuting with the requisite colors without sewing. Heretofore the stars of the union have been cut from cotton cloth and sewed to bunting, and the stripes cut from the wide bunting and stitched together. About two years ago, at the suggestion of a gentleman occupying a position at the head of one of the principal bureaus of the government, the manager of this company undertook to manufacture a flag having the stars and stripes fully and firmly shown without the aid of sewing. After many attempts and failures, and stimulated by the encouragement of many officers high in rank in both the army and navy, he has succeeded in producing flags of great beauty, durability, and extreme lightness. The manufacture of pendants, signals, and banners is soon to be added. This company has been from the commencement, and still is, under the management of Mr. D. W. C. Farrington, who is also agent of the company; and to Lowell belongs the honor of having inaugurated this patriotic and praiseworthy branch of industry in the United States. This company is also erecting machinery for the purpose of developing the manufacture of damasks, moreens, and other worsted fabrics.

In the same locality are also the works of the Wamesit Worsted Company, which is engaged in the manufacture of fine worsted yarns, consuming about two hundred thousand pounds of wool annually; also those of the United States Cartridge Company, which have recently been established for the manufacture of all kinds of fixed ammunition for breech-loading guns. These two last companies were organized in 1868, and are under the management of Mr. D. W. C. Farrington, who is the agent for all these works.

There is also an individual enterprise at this place entitled to particular notice. A stone shop has been erected and furnished with the best machinery and very superior mechanical skill, devoted to experiments in the improvement of fire-arms. This business is under the supervision of Capt. J. V. Meigs, a Tennesseean, and one of the most remarkable inventors of this prolific age. Heretofore it has been the most difficult study and problem with inventors to accelerate mechanical motion so as to approximate to the rapidity of the ignition of gunpowder, and with what success is well known. Capt. Meigs has, however, adopted a different policy. He has first increased the rapidity of ignition to the extreme bound of perfect safety by delivering the blow at the point most likely to produce instantaneous and complete combustion of the whole charge, and then, by his marvellous skill at mechanical combination, adjusted the speed of the explosive motion of the lock properly to meet the demand for rapidity of discharge, and its terribly fatal efficiency is exhibited by the possibility of the astonishing rapidity of fifty shots in nineteen seconds, still preserving its strength, durability, cheapness, effectiveness, and comfort in handling. Compared with this arm the famous needle gun is but a cross-bow; and, paradoxical as it may appear, this weapon of destruction must prove a life-preserver, as any struggle of arms, after the adoption of this gun, must be short, sharp, and decisive. Capt. Meigs has invented two guns, one new and the other an alteration, both of which have received the approbation of military committees and gentlemen, as well as gold medals and other awards for superiority. In the perfection of this arm it was found that ordinary ammunition was unsafe, improperly made, and inefficient; and to remove this difficulty he invented new cartridges, which combine every desirable quality; and for their manufacture, and all other kinds of fixed ammunition, the United States Cartridge Company was established. Charles K. Farmer, Esq., formerly of the Springfield (Mass.) Armory, is superintendent. Capt. Meigs has also brought out other important inventions; indeed, the fertility of his genius seems inexhaustible. Original minds, discoverers, and inventors have always been honored and esteemed; and, while these distinctions await Capt. Meigs, it is but just to remark that his genial qualities, his fund of generosity and humor, and all the elements of a distinguished and high-toned gentleman, mark him as eminently entitled to the consideration of his fellow-countrymen.

Some of the other enterprises around these falls are as follows: American Bolt Company, S. Hope, R. H. Butcher, and J. Winter, capital one hundred thousand dollars, consume two thousand tons iron and produce five million bolts per year; employ one hundred hands. Belvidere Woollen Company, capital one hundred thousand dollars, plain and twilled and fancy goods, product seven hundred and fifty thousand yards per year; employ eighty hands. S. W. Faulkner's Mills, capital one hundred thousand dollars, produce two hundred and forty thousand yards plain and twilled fancy goods and cassimeres, and fifty thousand yards fine Cashmeres, use two hundred thousand pounds wool per year. John Walsh, fine worsted yarns, one thousand spindles, use two hundred and forty pounds wool daily; thirty hands. S. R. Brackett, fine worsted, two hundred and sixty pounds wool daily, one thousand four hundred spindles; thirty hands.

Many years since, Mr. Nathan Ames established business at Massic Falls, which was afterwards carried on by the firm of Ames & Fisher. The sons of Mr. Ames have since become celebrated as manufacturers of arms at Springfield. Mr. Artemas Young followed Ames & Fisher, and built a brick mill on the site of the old Ames shop for manufacturing cloth; but failing in the completion of this enterprise, Mr. P. O. Richmond obtained possession, and established batting and paper mills, which he carried on successfully many years. The brick mill is now operated by Mr. William Walker, who manufactures woollens. In 1834, a shop was built on the island near the dam at Massic Falls. This building was occupied by George & Ephraim Crosby, but, being destroyed by fire not long after its occupation, it was not rebuilt. Previous to the year 1813 there was a saw-mill, owned by one Mr. Tyler, on the site of the mill which was erected in that year by Messrs. Phineas Whiting and Josiah Fletcher, who also built the dam the same year at this place, which is known as Wamesit Falls. Previous to this period that part of Belvidere lying between Fayette Street and the Concord River was an island, and there are those still living who have, floating in their light canoe, circumnavigated this now busy territory. In 1818, Messrs. Whiting & Fletcher sold and conveyed all their right, title, and interest to and in the water-power and adjacent lands to an enterprising gentleman from Charlestown, Mass., Mr. Thomas Hurd, who converted this mill into a woollen factory, and produced the first satinet goods which were woven by the power-loom in this country.

Mr. Hurd was one of those sagacious and scheming men who seek, and whose keen sense discovers and appreciates an advantage, while other men rub their eyes and listlessly wonder what he is doing. Not satisfied with the control of this power, and seeing the advantage of possessing the great power at the Pawtucket Falls, he purchased lands on either side of the latter, and put a grindstone, or some other simple machinery driven by a water-wheel, in motion, to establish his right to the privilege, and in 1826 put up a mill at the foot of the falls, apparently with the design to manufacture yarn or cloth; but his woollen mill on the Concord River having been destroyed by fire the same year, the mill just erected at the foot of Pawtucket Falls was taken down and rebuilt on the site of the one which was burned. A portion of the old foundation of this mill is still standing (1868) at the foot of Pawtucket Falls, — an interesting relic of the enterprise of one of those men who have left an honorable record in the history of Lowell. In the march of improvement this mill in its turn gave way to a substantial brick mill, built by the Middlesex Company on its site. Mr. Hurd also built and operated a brick mill. obtaining power from the Concord River by means of a canal, which followed the course of Warren Street, a little to the right of it, but finding there was a scarcity of water in dry seasons in the Concord River, a canal was dug from the foot of the Hamilton Canal, crossing Central Street, and passing to the rear and south side of Hurd Street, discharged its waters into the other canal between the two mills, thus securing a full supply of water at all seasons. Eventually Mr. Hurd's interest in the water-power at the Pawtucket Falls was disposed of to the Lock and Canal Corporation, when this auxiliary canal, which drew from the Merrimack, was discontinued and filled up. This latter canal will be remembered by many old residents of Lowell as the means and place selected by a Mr. Callender and Mr. Whipple voluntarily and most conveniently to "shuffle off this mortal coil." In 1821, Mr. Hurd sold the land and a water privilege on the east side of Concord River, comprising what is now called Belvidere Village (the island), to Mr. Winthrop Howe, who built a mill, and disposed of the surplus power to other parties, and quite a little cluster of shops eventually started up and are now in active operation in the vicinity. A flannel-mill, belonging to the Belvidere Woollen Company, now occupies the site, and is operated by the same power as the Howe mill, which was destroyed by fire some years since.

Prior to the construction of the Lawrence dam, in the glorious days when the salmon and shad had an unobstructed passage up the Merrimack, there was famous shad fishing from the foot of Wamesit Falls to the mouth of the river, — a gentleman still living having seen Capt. Whiting on one occasion make a miraculous haul. The seine was hauled on the spot directly between where Nos. 2 and 3, Middlesex Mills, now stand, and the fare was so large that boat-loads were taken from the net before it could be hauled in. A large raft was then constructed and boarded up several feet high at the sides, and the shad were placed in it and shipped for Newburyport; but, alas for human calculations! the raft grounded on an obtrusive boulder while running Hunt's Falls, and each individual item of the large cargo returned to its native element. The territory adjoining this fall, and in its immediate vicinity, mainly on the east side of the Concord, extending along that river to its confluence with the Merrimack and down the latter for a considerable distance, was the recognized home or head-quarters of the tribe whose name it still bears. Around these falls, on the very spot where mills and mechanics' shops now daily resound with the pleasant hum and din of industry, once stood their villages. A feeble tribe, and not a cannon-shot distant from their Pawtucket neighbors, they yet lived in security and peace. Under the immediate control of a subordinate chief, they nevertheless depended upon the Pennacook confederacy, to which they belonged, for protection, and leaned on the strong arm of Passaconnaway, their Bashaba, for support, and after him Wonnalancet, whom, after the shameless outrages perpetrated upon them by some evil-disposed people of Chelmsford, and being reduced to a mere remnant of a few dozen poor, forlorn Indians, they followed to the north country, and thus finally and forever disappeared from the Concord River. It is claimed by those who should be the best authority that Passaconnaway lived to a very advanced age, many years after his son assumed his duties and position. Gookin declares he saw him at Pawtucket when he had reached the patriarchal age of one hundred and twenty years; and, in "Tales of the Indians," he is represented, in 1682, as being at Agamenticus, in Maine, — identical with Aspinquid, — several years after Gookin wrote. Here it was that Gookin, to further the ends of justice, in those primitive days held his court, trying causes and meting out "equal and exact" justice with dignity and gravity. Here, too, the pious Elliot, in his little log chapel, preached "peace on earth and good-will to man," while stalwart savages, steeped to the eyebrows in Paganism, listened with unction, if not compunction, and stupid decorum, shrugging their shoulders as they retired at the disparity of the preaching and practices of the pale-face.

The Merrimack River, at Lowell, is spanned by two bridges, — one at Pawtucket Falls, the other leading to Central Village, which formerly belonged to the town of Dracut, but was annexed to Lowell some twenty years since. This last bridge was rebuilt a few years ago at a cost of about thirty-five thousand dollars.

CHAPTER XI.

Pentucket Navigation Company. — Nicholas G. Norcross. — Andover. — Methuen. — The Spicket. — Lawrence. — History of its Manufactures. — Bradford. — Little River. — Haverhill. — The Pow-Wow. — Amesbury. — History of its Manufactures. — Newbury. — Salisbury. — Newburyport. — Plum Island. — Seabrook. — Conclusion.

THE improvements made in the channel of the Merrimack to facilitate its navigation have been extensive, but mostly accomplished by individual enterprise. Many business men, who have required the river as a means of transit for heavy freight, lumber, rafts, and boats, have been obliged, at first, to make a heavy outlay for this purpose. Many gentlemen, knowing the great importance of this stream as a thoroughfare and means of transportation, have endeavored to secure organized and systematic action with larger means; but it was not until John Nesmith, Esq., of Lowell, who was born by the side of the Merrimack, and who, during a long and successful business career, has been intimately connected with it, — not until he took hold of the project was much of real progress made. In 1867, John Nesmith and others petitioned the Legislature, and were incorporated as the Pentucket Navigation Improvement Company. The object of this company was to provide a cheap and convenient outlet for accumulating freights of merchandise and perhaps to reduce the burdensome monopoly enjoyed by the railroads. To accomplish this purpose, or to guarantee any chance of success, it had been found necessary to unite and put forth the efforts and means which nothing less than a heavy corporation could furnish. The Legislature granted a charter the year mentioned, but it was absurdly hampered with such restrictions and limitations as had the effect to render the possession of the charter of no use whatever. The United States Government, it was confidently expected, would make appropriations for tide-water improvements, while from that point the company should make the river navigable for light-draught boats to the important city of Lowell, in the whole a distance of more than thirty miles. The charter required that the company should push its business so rapidly that vessels of twenty or more tons could reach Hunt's Falls, near Lowell, at the ordinary stages of the water, within three years. The company, foreseeing that it could not be done, did not make the attempt, but has confined its efforts to securing the removal of these legislative restrictions. Should this, very properly, be done, the company will then be in condition to accomplish a work of great value to all whose business is much with or on the river. The current of the Merrimack is so rapid and strong as to necessitate a continual and extensive outlay of labor and capital to maintain an available ship channel. Added to this, much of the business done on the river has the effect to increase its natural tendency to obstruct or obliterate any artificial channel. Its annual, often semi-annual, rise of twenty or more feet, the snags and driftwood brought down by these resistless floods, the constant fretting of the banks, the washing in of vast deposits, the constant plying of heavy boats, barges, and rafts, and the annual "drive" of many millions feet of saw-logs, all conspire to affect the channel more or less. Of this last great business, the lumber trade, a brief historical sketch cannot fail to be interesting.

In 1844, Nicholas G. Norcross, who had already made himself rich, and earned the title of "The New England Timber King," on the Penobscot, came to Lowell and established himself permanently on the Merrimack. Prior to that time timber had been brought down in rafts, locking around the falls; altogether a slow and tedious process. Mr. Norcross prefaced his operations by the outlay of more than one hundred thousand dollars in improving the channel and adapting it to his purposes. He blasted rocks and removed obstructions, bought land and provided for the stringing of booms for timber harbors, bought rights on some of the important falls, built two dams on the Pemigewasset at Woodstock, and purchased the Elkins grant of eighty thousand acres of heavy timber adjoining the above-named town, Lincoln, and several others. He also bought a tract of forty thousand acres in the ungranted lands of New Hampshire, and several other tracts. Having prepared the river to receive the logs and for the "drive," the mode of proceeding was to repair to the timber forests with a force of choppers, some one hundred and fifty or two hundred men, cut and haul the logs for the coming "drive," which usually commenced about the middle of April; the annual "drive" varied from eight to fourteen millions feet. A store was opened by the manager, from which the commissary of the force drew all supplies. Large, warm log barns were erected to shelter the large number of oxen used, and hay and feed were purchased of the farmers in the nearest settlements. The logs afloat, and the woodsmen became amphibious and took to the river. As the logs send off, the river becomes a scene of picturesque beauty, and as it is with humanity so it is with these logs, — they cannot all "head the procession;" some strike a more rapid current than others, some meet with obstructions by which their advance is retarded or suspended, but the van moves steadily and rapidly on, and soon the river surface is thickly dotted with logs for a distance of from thirty to fifty miles. The force is divided into "gangs," or "teams," and, under the lead of experienced river-men, proceed to the principal falls and rapids, where the chief difficulties of the drift are generally experienced, a small force following, as a rear-guard, to push off any stragglers that may have been crowded ashore or grounded on some unseen obstruction. "Running the falls" is a wild, exciting, and very interesting spectacle, lasting many days. Sweeping towards the head of the falls like a vast host in solid column, or like an Alpine avalanche, they plunge down the roaring, boiling, seething rapids in furious, headlong haste; the narrow gorges between the projecting ledges are choked with them; they are thrown over and across each other, sometimes assuming a perpendicular attitude and falling with a tremendous crash and plash. Sometimes a single log gets arrested by the outcropping ledge, and, being securely held, becomes a literal stumbling-block to others, and soon a huge "jam" is collected and tightly held, — a thousand logs, more or less, piled up without regard to system, as though prepared for a huge bonfire. Now is the time when the skill and daring of the "drivers" become conspicuous; with the agility of squirrels they cross the rapids on the moving mass and gain the pile which has been crowded out to cut it adrift. Experience points out the stick which holds them fast and prevents the whole from moving on, and not unfrequently, by a vigorous push with the levers or pike-poles in the right place, or a few strokes of the axe, the "jam" breaks loose, again falls into line, and moves rapidly on. Each man takes to a log, and navigates to a place of safety, not hugging it after the manner of embracing a friendly lamp-post, but standing erect, and, if the logs are rolled by the undertow, they retain their position upon the highest exposed convexity by moving their feet, and thus, while they easily maintain a position on the exposed surface, offer no objection to its revolution. The view of the men in active duty upon any of the great falls on the river is picturesque and interesting. Moving actively about in a uniform of gray and red, jumping from log to log, or boldly floating through the turbulent waters on this narrow craft, the river everywhere filled with moving timber, running the falls is a picture worth looking upon.

In 1845, Mr. Norcross built a large lumber-mill at Lowell, where, with "gangs" of saws, upright and circular, he wrought out much of the lumber for the mills and the dwellings of the city. This mill was twice destroyed by fire, but was soon rebuilt. He also built a large mill at Lawrence, which was managed by his brother, J. W. Norcross. Mr. Norcross died in 1860, since which time the business has been conducted by I. W. Norcross, Charles W. Saunders, and N. W. Norcross. While engaged in the lumber business a financial crisis, such as business men often experience, overtook Mr. Norcross, and, with a view to the continued prosecution of the trade, a company was formed, consisting of N. G. and I. W. Norcross, John Nesmith, Abner Buttrick, H. Pillsbury, William Fiske, and others, called the Merrimack River Lumber Company; but the management of the lumber trade seems to have again reverted to the original hands, and is now conducted by Norcross & Saunders.

Mr. Norcross was a remarkably energetic business man. Quick to see, and prompt to decide and act, his years were crowded with stirring events. In the prosecution of great enterprises like this on the Merrimack, he had the sagacity to comprehend the situation, and by a liberal investment of labor and capital laid a substantial foundation for ultimate success, which is its surest guaranty.

With frequent high water, a strong and rapid current, the strong force of pressure and displacement occasioned by the plying of river craft, the conveyance of the immense lumber drives, and many other causes, it would seem very clear that the Pentucket Company, with the proper legislative encouragement, would be an untold advantage to the business of the Merrimack River.

Andover was settled in 1643, and incorporated five years later. It is watered by the Shawsheen River, a tributary of the Merrimack, having its confluence on the right bank at Lawrence, and there are excellent facilities for manufacturing, which are extensively improved. There is a celebrated Theological Institution, and other institutions for educational purposes of a high character located here.

Methuen is situated on the north bank of the Merrimack, and is an extensive manufacturing and mechanical town. The Spicket River, which has its source in Hampstead, N. H., passes through this place, and having a splendid fall of thirty feet, furnishes a fine water-power. Methuen was detached from Haverhill in 1725, and has maintained a steady growth and prosperity, and takes high rank among the flourishing towns of Essex County. The Spicket empties into the Merrimack, on its left bank, at Lawrence.

Bradford was incorporated in 1675. It is a very pleasant town, the land rising gradually from the Merrimack, culminating in elevated ridges and gentle hills, and as the neat dwellings are built along the slopes, the appearance of the town from a distance, or from some neighboring elevation, with the broad and placid Merrimack in the foreground, is picturesque and pleasant. The shoe business is the principal interest of Bradford; in fact, it may be said it is only a repetition of its larger neighbor on the opposite side of the river. Besides the Boston & Maine Railroad Bridge there is what is known as the Toll Bridge, which was made free in July, 1868, by indemnifying the stockholders, and the Rock Bridge, so called, further down the river.

"The Agawam tribe occupied the eastern part of what is now Essex County, in Massachusetts, extending from tide-water upon the Merrimack, round to Cape Ann. Their territory skirted upon two sides by the Merrimack and Atlantic, indented by bays, intersected by rivers, and interspersed with ponds, was appropriately called *Wonnesquamsauke*, meaning, literally, "The Pleasant Water Place;" the word being a compound from *wonne* (pleasant), *asquam* (water), and *auke* (a place). This word was sometimes contracted to Wonnesquam, often to Squamsaukee, and still oftener to Squam or Asquam. The deep, guttural pronunciation of *Asquam* by the Indians sounded to the English like *Agawam*, and hence the word as applied to the Indians of that locality. Several localities in Essex County are now known by names contracted and derived from this Indian word, *Wonnesquamsauke;* as "Squam," the name of a pleasant harbor and village upon the north side of Cape Ann, and 'Swamscott,' the name of a pleasant village in the eastern part of Lynn." *

The city of Lawrence, the ancient seat of the Agawams, is another of those magnificent and forcible illustrations of the almost incredible power and capacity of the Merrimack. For two centuries the river and the land literally ran to waste; but sparsely settled, in productiveness meagrely requiting the tiller's industry, it seemed destined, like the other points of manufacturing interest along the Merrimack, to a career of barrenness and comparative worthlessness, until the splendid water-power caught the eye of the sagacious manufacturer, when a change, rapid as wonderful, came over the scene: the desert waste grew green, active, busy life dispelled the unpleasant silence, and the solitary place forthwith resounded with the cheerful rattle of machinery, the ring of the anvil, the vigorous strokes of the artisan and mechanic, the whirl and bustle of trade, and the constant rush of steadily augmenting throngs where once the few hardy fishermen, along the falls, at the mouth of the Shawsheen and the Spicket, captured the magnificent and delicious salmon, the bony shad, and the slimy, squirming eel, a change is wrought, sudden and complete. Monster factories, and a beautiful city of elegant public buildings, handsome, convenient, and comfortable dwellings, workshops, school-houses, and churches, a large, enterprising and industrious community, now adorn, enliven, and beautify the place. As early as 1835, Hon. Josiah G. Abbott, and other gentlemen interested in manufacturing, examined this water-power with a view to the establishment of extensive works: but it was not until ten years later that active and decisive steps were taken which have resulted in the springing up, as if by a touch of the enchanter's wand, of this extensive manufacturing city.

About 1835, the enterprising people of Methuen discussed the project of turning the Merrimack River into the Spicket, as there was a fine fall of thirty feet on the latter stream, but an insufficiency of water for very extensive operations. Surveys were made

* Potter's History of Manchester.

by Stephen Barker, and it was ascertained that the scheme was not feasible. From that time, however, attention was directed to the falls in the Merrimack, and in 1844, the proprietors of the land and water power were incorporated as the Essex Company, capital, one million dollars, and steps were at once taken for building up a large manufacturing city. In 1845, the first boarding-house was erected, and the construction of the dam begun the following year. Same year the "Bay State," Atlantic, Cotton, and Union Mills, also Bleachery and Dyeing Company. The total capital of these corporations was four million five hundred thousand dollars. In 1852, the Pacific Mills Company was incorporated; capital, two million five hundred thousand dollars. In 1853, the Lawrence Duck; capital, one hundred and twenty thousand dollars; and also, this year, the city of Lawrence was chartered, this name being given it in compliment to Hon. Abbott Lawrence, who was heavily engaged in manufacturing in New England, and who was for several years envoy to England. In 1854, the Pemberton Mill was incorporated. This mill fell in 1859, and was rebuilt the same year; capital, four hundred and fifty thousand dollars. Everett Mill, incorporated 1860; capital, eight hundred thousand dollars.

The name of the Bay State Company has been changed to Washington Mills; capital, one million six hundred and fifty thousand dollars. The Arlington Woollen Mills Company was incorporated in 1865; capital, two hundred thousand dollars. There are two fine bridges at Lawrence, the railway and highway.

On the 10th of January, 1860, at a few minutes before five o'clock, P. M., an appalling catastrophe occurred in this city. The Pemberton Mill, in which were employed more than seven hundred operatives, suddenly fell, burying all of them amid the wreck, and, to add, if possible, to the consternation and horror of the scene, the ruins took fire. About one hundred lost their lives, and it was estimated that a majority of those in the mill were either killed, or more or less severely injured, either by the fall or the fire. By the census of 1865, the population of the city was twenty-one thousand seven hundred and thirty-three. The fall at Lawrence is twenty-six feet, and the fall in the river, from the dam to tide-water, eight miles, is eight feet only.

Around these falls the Agawams often congregated, and ranged along the Merrimack to the sea. Fishing the numerous small streams, which, rising in New Hampshire, sweep down through Haverhill, Amesbury, Salisbury, and Seabrook, crowded with alewives; then again, exploring the bays, inlets, and small rivers, around the eastern shore, from Newburyport to Nahant; swarming across the gentle hills and wooded plains, and swampy lowlands of Essex, they lighted their camp-fires around interior ponds, well stocked with pickerel. Here, with pomp, parade, and bright prospects, went to reside the Princess of the House of Pennacook. Winnepurket, often called George, Sachem of Saugus, sued for, and received the hand of the daughter of Passaconnaway in marriage, and soon after, on a nice point of barbarian punctilio, a fierce embroglio, involving the estrangement of the young couple, arose between father-in-law and son-in-law, which, only for the timely advent of the pale-faces, might have resulted in a rebellion, large bounties, and an interminable firebrand or sapling dance. "The poet of the Merrimack" has chronicled this emeute in immortal verse,* Winnepurket being characterized as "dog of the marshes," among other maledictions heaped upon his devoted head by his incensed and implacable father-in-law. Winnepurket married again,† and at the close of Philip's war, he, together with several cargoes of Indians, was sent to Barbadoes, and sold into slavery, — a trans-

* " I bore her, as became a chieftain's daughter,
Up to her home beside the flowing water.

" If now, no more for her a mat is found,
Of all which line her father's wigwam round,
Let Pennacook call out his warrior train,
And send her back with wampum gifts again.

"' Dog of the marsh!' cried Pennacook, ' no more
Shall child of mine sit on his wigwam floor.
Go! let him seek some meaner squaw to spread
The stolen bear-skin of his beggar's bed.

"' Son of a fish hawk! let him dig his clams
For some vile daughter of the Agawams,
Or coward Nipmucks! May his scalp dry black
In Mohawk smoke, before I send her back.'"

† This second wife bore the no doubt euphonious, but unpronounceable, name of Ahawayetsquaine.

action sternly rebuked by their more moral (?) descendants. He finally returned, and died at the age of sixty-eight.

Soon after leaving Lawrence, the river shows the effect of tide-water very distinctly, its current being less perceptible, moving as a compact body rather than a collection of buoyant particles, sluggish and darker than when rippling and bounding over its rocky and descending bed. Notwithstanding the river has very nearly found its final level, the banks are steep, almost to the ocean, and through Haverhill and Bradford, which are separated by it, retire boldly to a considerable height, affording building sites at once picturesque and pleasant, rising one above another, like tiers of parquette boxes, giving to all an unobstructed view of the grand panorama of river and landscape. Haverhill is thus laid out, a large, handsome, healthy village. It was formerly called Pentucket, and was settled in the year 1640, by William White, Samuel Gile, James Davis, Henry Palmer, John Robinson, Christopher Massey, John Williams, Richard Littlehale, Abraham Tyler, Daniel Ladd, Joseph Merrie, and Job Clement; the last four from Ipswich. It was the thirtieth town settled within the present limits of Massachusetts, and was the forty-ninth town settled in New England, and was the thirty-second incorporated town in the State.*

The deed of the town given by the Indians — the original instrument — is still in the possession of the town of Haverhill.

On the 5th of March, 1643, there was a severe shock of an earthquake experienced here, which alarmed all classes of people greatly, but did no serious damage.

July 5th, of the same year, a very violent hurricane passed over this section of country, prostrating multitudes of trees in this town, and killing one Indian. In Newbury, it raised the church from its foundation, and, though the people were assembled in it, no one, it is believed, was injured.

Kenoza, or Pickerel Lake, is the largest body of water in Haverhill. Its area is three hundred acres, and its surface is one hundred and fifty feet above the Merrimack. Great Hill, so called, is the highest elevation of land in the town, and the second highest in Essex County. It is situated a mile north of the above-named lake, and is three hundred and forty feet above the sea.

* Palfrey.

Little River, a lively but inconsiderable affluent of the Merrimack, has its source in Plaistow, N. H., and effects a junction with the Merrimack here. Hale's Flannel Mills are located on this stream, as also numerous shops and mills of various kinds along the entire course of the river and its tributaries.

Fishing River, so called from the quantities of alewives formerly taken, is the principal tributary of Little River.

Meadow River, which also runs through Haverhill, rises in Newton, N. H., and drives many small mills, shops, etc.

There are two long bridges spanning the Merrimack, — the Boston and Maine Railroad bridge, and a toll bridge, which is the highway between this town and Bradford.

Haverhill supports eleven churches, an excellent and flourishing high school, and a very good library. Its town hall is a costly, elegant, and convenient structure. There are nearly thirteen hundred houses, and one hundred shoe manufactories, which is the principal business of the place. The population is largely transient and floating, varying as much as four thousand between high and low tide of business.

" The view from Silver Hill is exceedingly beautiful. Before us, and almost at our very feet, lies the pleasant village of Haverhill, with its twelve hundred dwelling-houses, its one hundred shoe manufactories, and its eleven churches. Its natural situation is uncommonly fine. Built upon a gentle acclivity, the houses rise one above another in such regular order that nearly every one can be counted. The Merrimack, dotted here and there with a variety of craft, from the light and trembling skiff to the heavy gondola, and the still more imposing and majestic moving ocean craft, with their broad, white sails and tall masts overshadowing the water, and, spanned with its bridges, flows calmly at its base, not in straight, monotonous course, but with a gentle meandering, of which the eye can never tire.

" Across the river are seen the smoothly rounded hills, the green and fertile fields, and the pleasant villages of Bradford and Groveland. To the south rise the hills of Andover, with their wooded slopes dotted here and there with neat, white farm-houses. A little to the west, the tall spires, just peeping above the hills, point out the whereabouts of the city which sprang into existence almost like Jonah's gourd, — the city of Lawrence. A little farther still to the west, and the same signs indicate the spot long ago settled by the hardy sons of Haverhill, the village of Methuen. In the dim distance beyond, enveloped in misty blue, can be traced the outline of Mount Wachusett. Still farther towards the west, as if it were not well the eye should roam too far, the Scotland and West Meadow Hills shut out the more distant view beyond, but not until we have caught sight of the tall peak of the grand Monadnock. Sweeping towards the north we have a view of the thrifty farms of the West Parish, with the granite hills of New Hampshire in the background.

"To the north, the eye rests upon a fine succession of green fields and wooded slopes, marking a section of the town which suffered the most severely from the atrocities of the murderous savages. There the brave and resolute Hannah Bradley was twice taken captive; there the lion-hearted Hannah Dustin was captured, but not conquered, and there stands her monument; there the heroic Thomas Dustin defied the murderous tomahawk to harm the humblest of his little flock. There, too, upon that gentle slope, the brave Captain Ayer and his little band boldly attacked the retreating foe upon the memorable 29th of August, 1708. From this summit might have been heard the warwhoop, and have been seen the gleaming tomahawk, in nearly every attack made upon the inhabitants of Haverhill by the savages.

"The valley of the Little River, or Indian River as it was also once called, of which the section just mentioned forms a part, is here seen in all its beauty, as it stretches, with its charming succession of hill and dale and meadow, from the Merrimack far back among the granite hills of our sister State. This view alone is well worth a visit to the broad summit of Silver Hill." *

Amesbury and Salisbury are so intimately and naturally connected, so interwoven in business relations, that it is impossible to obtain exactly and separately the business of each. They were originally one town, and were settled by Simon Bradstreet. Daniel Dennison, and others, to whom it was granted as a plantation, under the name of Merrimack. The following year it was incorporated under the name of Colchester, and in 1640, by direction of the General Court, it took the name of Salisbury. In 1668, the town was divided by the course of the Pow-Wow, the boundary between the towns, and the

* Chase's Haverhill.

dissevered portion lying on the north side or left bank of the Merrimack took the name of Amesbury.

The Pow-Wow River, a considerable tributary of the Merrimack, takes its rise in Kingston, N. H., and effects a junction with the Merrimack in this town, and affords one of the best manufacturing privileges to be found in the country. Like most other New England streams, the first use made of this was for saw and grain mills, and its usefulness was limited to these and kindred purposes. Nail and iron works followed.

In the year 1812 the manufacture of cloth was begun, and the first contract for clothing our troops in the war which began the same year was filled by this mill. From that time to the present, manufacturing has been steadily increasing, and it has reached a magnitude not probably dreamed of by its original projectors.

Previous to the year 1854 there were two corporations, the Salisbury Mills and the Amesbury Flannel Mills. About that time these were consolidated under the name of the Salisbury Manufacturing Company. In 1856, the present company purchased all the property and effects of this concern, and was incorporated by the name of the Salisbury Mills Company, with a capital of seven hundred and fifty thousand dollars, which has since been increased to one million dollars.

This company has, at the present time, ten large woollen mills, containing seventy-five sets of machinery, and two thousand cotton spindles. These mills are located on the Pow-Wow, and obtain their motive-power from a total aggregate fall of seventy-five feet, which is wholly included within a distance of seventy rods. This company own the entire privilege of this river, from its source in New Hampshire to its confluence with the Merrimack, which occurs a half a mile below the mills, and flows more than three thousand acres, affording a never-failing reservoir. The water is used over five different times, exclusive of the times which it is used in running three saw and three grain mills, which the company own and operate. Almost every variety and description of woollen fabrics are manufactured here, and the enormous quantity of more than seven and a half miles in length of manufactured goods is produced daily, — a business amounting to over three millions of dollars per annum. There are five feet of tide-water, affording an easy, cheap, and convenient conveyance of heavy articles, such as machinery, coal, wool, etc., to and from the mills. These mills use the almost fabulous quantity of between four and five millions pounds of wool per annum.

Mr. Samuel H. Shepard, who has charge of the wool department, met with a frightful accident and a very narrow escape from death, in one of the large wool houses in Boston. He fell a distance, said to be fifty-four feet, with no injury except a compound fracture of one ankle.

John Gardner, Esq., is treasurer of this company, he having held that position since 1856. The present agent, who has charge of these extensive works, is M. D. F. Steere, who has occupied his present position since the spring of 1858.

There is also a wool hat manufactory, with a capital of one hundred and fifty thousand dollars, with three mills, each of which is a complete factory of itself, taking the raw material and turning out hats finished and trimmed, ready for retail. This company employs two hundred and fifty hands, and the daily product is two hundred and twenty-five dozen elegant hats. In addition to the foregoing, there are other mills and works with a united capital of nearly two hundred thousand dollars.

Carriage-making is carried on here to an extent exceeding any other place in the State, in numbers and quality, and Amesbury vehicles enjoy a high, extensive, and well-merited reputation. Ship-building is also carried on, some half dozen, of various tonnage and of superior material and model, being annually launched.

Pow-Wow Hill, an elevation of three hundred feet, is said to have been the council-chamber of the aboriginal settlers, from whence they could survey the surrounding country and gaze far out upon the blue waste of waters.

In 1643, under a false construction of her boundary claim, Massachusetts seized upon Hampton, Exeter, Portsmouth, and Dover, in New Hampshire, uniting with them the towns of Salisbury and Haverhill, in Massachusetts, formed a new county, which was called Norfolk, Salisbury being the shire town, which it continued to be until 1679, when the boundary was readjusted, and the four towns first named reverted to their original and rightful jurisdiction.

In August, 1737, commissioners appointed by the crown assembled at Hampton Falls to settle the controversy. On this occasion, the

General Court of New Hampshire convened at Hampton, while Salisbury had the distinguished honor of being the place of meeting of the great and General Court of Massachusetts.

As the Legislature of New Hampshire was to meet on the 4th day of August at Hampton, that of Massachusetts adjourned at Boston to meet again on the 10th of the same month at Salisbury, the adjoining town. Much preparation had been made in Boston for this transient session of the portable "great and General Court," and Gov. Belcher, in company with many dignitaries, rode in great pomp and state, escorted by troopers and horsemen, to Newbury, from whence he was escorted by additional companies to the George Tavern, at Hampton Falls, when, as it was long before the enlightened age of Maine laws and State constables, it is very probable the governor and his friend took something — to eat.

The governor and his *cortége* made up a brilliant pageant for that age, much to the disgust of the admirers of official simplicity of manners, and the show was lampooned severely. Here is a sample: —

> "Dear Paddy, you ne'er did behold such a sight,
> As yesterday morning was seen before night.
> You in all your born days saw, nor I didn't neither,
> So many fine horses and men ride together.
> At the head, the lower house rode two in a row,
> Then all the higher house trotted after the low;
> Then the governor's coach galloped on like the wind,
> And the last that came foremost were troopers behind.
> But I fear it means no good to your neck or mine,
> For they say 'tis to fix a right place for the line."

All this parade amounted to nothing, except it was to furnish a "Court Record" item for the "Boston News Letter," thus: —

"HAMPTON FALLS, in New Hampshire, Aug. 18.

"On Monday last, at eight o'clock in the morning, His Excellency, our Governor, attended by several of His Majesty's Council and sundry other gentlemen, set out for Londonderry, and on Monday night lodged at the house of Robert Boyes, Esq., in that town. On Tuesday His Excellency went to Amoskeag and returned in the evening to Mr. Boyes's, and yesterday came back to this place in good health, having dined on his way hither with Mr. Sanborn, of Kingston (the Representative from that town). His Excellency was much pleased with the fine soil of Chester, the extraordinary improvements at Derry, and the mighty falls at 'Skeag."

Salisbury beach is considered remarkably beautiful, and is much frequented.

West Newbury was taken from the ancient town of Newbury in 1819. It is a good farming town on the south side of the Merrimack. Marble and a variety of hornblende or *amianthus*, the finer varieties of which have been manufactured into cloth which will not burn, and a mineral sometimes called asbestos is quarried here. There are also carriage manufactories, but the principal business is professional farming, carried on by gentlemen of means, who realize twenty per cent. of pleasure to one of profit. There is on the highway on the Merrimack a bridge known as the "Chain Bridge."

"Ould Newberry," as it was anciently called, was settled in the spring of 1635. It derives its name from Newbury, a town in England; being so named by the wish of Rev. Thomas Parker, who was the first minister, and who had formerly preached in Newbury, England. The Indians called the place *Quascacunquen*, which signifies a "waterfall," — in this case the waterfall on Parker River. There were ninety grantees of the town, most of whom, in a small vessel, came round the coast, entered the Merrimack River, and landed where the city of Newburyport now stands. The first white child, Mary, daughter of Thomas Brown, was born in 1635. Joshua, son of Edward Woodman, was born the same year. The first named lived to the patriarchal age of eighty-one years.

In 1657, Thomas Macy was prosecuted for a violation of the law against harboring or entertaining Quakers. It appears that, during a violent shower of rain, several men sought shelter under his roof. They were strangers to him, but two of them proved to be William Robinson and Marmaduke Stephenson, who were afterwards hung in Boston for the heinous crime of being Quakers. These gentlemen inquired the way to Hampton, and in about three-fourths of an hour, the storm abating, went their way. For this act of anti-Puritanic hospitality, although he protested his entire ignorance of the persons or their religious character, and offered a humble apology for the grave (!) offence, Mr. Macy was fined thirty shillings. Being himself a Christian gentleman of the practical type, he was disgusted with the bigotry and intolerance exhibited, and, embarking his family and effects in an open boat, "left the country," and went to the island of Nantucket, where he passed the remainder of his days.

It may be gratifying to know that the statutes paid attention to personal adornment, as well as the religious views of the community. Thus, no person not possessed of the value of two hundred pounds was permitted to wear silks, and many women were fined for a violation of this law. So the two-hundred pound ladies looked down with commiseration, perhaps, on the poorer ones.

Newburyport, directly at the mouth of the Merrimack, is, probably, the smallest town in area in the United States, it being only one mile square. It was formerly a part of Newbury, and was separated from that ancient town in 1764. Formerly the foreign trade was extensively carried on, but has been impeded by the bar at the mouth of the river. Still the town is extensively engaged in fishing and freighting, besides some ship-building, cotton manufacturing, and many other kinds of business.

Josiah Bartlett, one of the signers of the Declaration of Independence, born in Amesbury, died in this place May 19, 1795. George Whitefield, one of the founders of Methodism, died in this town in 1770. He was born in Gloucester, England, December 16, 1714, and the inscription on his cenotaph says: " In a ministry of thirty-four years he preached more than eighteen thousand sermons, and crossed the Atlantic thirteen times. As a soldier of the cross, humble, devout, ardent, he put on the whole armor of God, preferring the honor of Christ to his own interest, repose, reputation. and life." George was of humble and obscure origin, the first and the last, of any note, who bore the name of Whitefield. He was a natural orator; and in him was concentrated that masterly and indefinable power of eloquence which sways the emotions, directs and controls the impulses, and captivates while it enlarges the mind and the heart. Possessed of a gorgeous imagery, which afterwards made his sermons so majestic, effective, and beautiful, he early looked to the stage, which, however, was a sphere too limited for his masterly powers and unbounded benevolence. His stage was to be the universe, his audience a sinful world, his reward reclaiming and restoring fallen man to Christ. He preached in New England to multitudes, who, spell-bound, were moulded to his will by the irresistible power of persuasive and glowing eloquence. Whenever and wherever he preached, shops were closed and secular business generally suspended; crowds followed to listen with almost unseemly eagerness, and the masses, enraptured by his fervid eloquence, were melted to tears by his sublime pathos, or inspired with an ecstasy of hitherto unfelt and unknown joy and peace and happiness, as he portrayed the munificence of a Saviour's undying love. Like all illustrious and conspicuous men, he had bitter and relentless enemies; such minds as prefer lowering the angels to elevating or raising mortals up; but whether eminent professors or others, the malignity of their enmity was inbred and not contagious, and the farthing rush they opposed to the effulgence of this brilliant and blazing meteor was totally eclipsed and consigned to oblivion by the inestimable superiority of his intellectual power and Christian character; by the power and success and magnitude of his religious efforts, and the unfading glory and renown justly due and generously accorded to him, and his cherished and revered memory, for the beneficent result of his sacrifices and labors in the cause to which he devoted his matchless eloquence, his best efforts, and even life itself.

Hon. Caleb Cushing was born at Salisbury, Mass., January 17, 1800; graduated at Harvard College in 1817; was elected to Congress in 1835, where he served with distinguished honor for several years. He was appointed to a seat in the cabinet of President Tyler, but was rejected by the Senate, and in 1843 was appointed Commissioner to China for the United States Government, being the first minister sent to the Celestial Empire. He was brigadier-general in the Mexican War, and was a member of President Pierce's cabinet, discharging the duties of his position with consummate ability. On entering the cabinet he resigned his position on the Supreme Bench. As a debater he is considered to be without an equal in New England. He is, personally, the most popular of men, and dispenses his large means with a liberal hand.

There are many manufactories and mechanical works in Newburyport, some of them of considerable magnitude. Among these may be mentioned the American Machine Company (paper collars), with a capital of eighty-two thousand dollars; Globe Steam Mills (print cloths), capital, two hundred thousand dollars; James Steam Mills (sheeting and shirting), capital, two hundred and fifty thousand dollars; Merrimack Arms Manufacturing Company (rifles, guns, and pistols), capital, two hundred thousand dollars; Ocean Steam Mills (print cloths and sheetings), capital, three hundred and sixty thousand dollars; and various other establishments.

"Plum Island, at the mouth of this river (Merrimack), to whose formation, perhaps, these very banks have sent their contribution, is a similar desert of drifting sand, of various colors, blown into graceful curves by the wind. It is a mere sand-bar exposed, stretching nine miles parallel to the coast, and, exclusive of the marsh on the inside, rarely more than half a mile wide. There are but half a dozen houses on it, and it is almost without a tree or a sod, or any green thing with which a countryman is familiar. The thin vegetation stands half buried in the sand as in drifting snow. The only shrub, the beach plum, which gives the island its name, grows but a few feet high; but this is so abundant that parties of a hundred at once come from the main land and down the Merrimack, in September, and pitch their tents, and gather the plums, which are good to eat raw and to preserve. The graceful and delicate beach pea, too, grows abundantly amid the sand; and several strange, moss-like, and succulent plants. The island for its whole length is scalloped into low hills, not more than twenty feet high, by the wind, and, excepting a faint trail on the edge of the marsh, is as trackless as Sahara. There are dreary bluffs of sand and valleys ploughed by the wind, where you might expect to discover the bones of a caravan. Schooners come from Boston to load with the sand for masons' uses, and in a few hours the wind obliterates all traces of their work. Yet you have only to dig a foot or two anywhere to come to *fresh water;* and you are surprised to learn that woodchucks abound here, and foxes are found, though you see not where they can burrow or hide themselves. I have walked down the whole length of its broad beach at low tide, at which time alone you can find a firm ground to walk on, and probably Massachusetts does not furnish a more grand and dreary walk. On the sea-side there are only a distant sail and a few coots to break the grand monotony. A solitary stake stuck up, or a sharper sand-hill than usual, is remarkable as a landmark for miles, while for music you hear only the ceaseless sound of the surf, and the dreary peep of the beach-birds." *

The settlement of Seabrook was commenced in 1638. It lies on the north side of Newburyport Bay, opposite the mouth of the Mer-

* Thoreau's "Week on the Concord and Merrimack Rivers."

rimack River, in the extreme south-eastern angle of New Hampshire. A portion of its present limits was formerly in Massachusetts, but was restored to the rightful jurisdiction of New Hampshire by the adjustment of the boundary line between the States, which occurred in 1741. The former line, from the "Bound Rock" at the mouth of the river, — on which can still be traced the inscription, "A. D., 1657, H. B.," — can yet be traced to a rock near the brick schoolhouse, marked "B. T." Seabrook was granted, June 3d, 1768, to Jonathan Weare and others. The Weare family have been among the most able and prominent of the early settlers of New Hampshire. Nathaniel Weare was an agent for the colonies, and was in England much of the time, urging attention to the complaints of the colonies against Edward Cranfield, the royal Governor. Nathaniel Weare, Jr., was much in public life, and was familiar with and skilled in the transaction of public business. Meshech Weare, son of the latter, was also a resident of this town, and was at one time President of New Hampshire, and was a prominent character as well as one of the ablest men of his time. Among the early settlers were the Felches, Christopher Hersey, Joseph Dow, Thomas Philbrick, and the Goves. These names are still familiar wherever there is a society of Friends in New Hampshire. Edward Gove was always bitterly opposed to the arbitrary and tyrannical policy of England. He was arrested, tried, and convicted of high treason, and was a prisoner in the tower of London for three years, when he was released and returned to his friends. The following is a copy of his pardon : —

"JAMES R. Where as Edward Gove was neare three years since apprehended, tryed & condemned for High Treason in our Colony of New England in America, and in June, 1683, was committed prisoner to the Tower of London, we have thought fit hereby to signify our Will and Pleasure to you that you cause him, the said Edward Gove, to be inserted in the next general pardon that shall come out for the poor Convicts of Newgate without any condition of transportation, he giving such security for his good behavior as you will think requisite. And for so doing this shall be your warrant.

"Given at our Court at Windsor the 14th day of September, 1685, in the first year of our reign. By his Maj. his command. Sunderland. To our trusty and wellbeloved the Recorder of our Citty of London and all others whom it may concern. Edward Gove to be inserted in ye General Pardon."

There is a letter still extant, written to him at this period by his daughter Hannah and her husband, Abraham Clements, the superscription being as follows: "For my honored father Edward Gove. In the tower or elsewhere I pray you deliver with care." The letter bears date as follows: "From Hampton the 31 of ye first month, 1686." The Quaker sect was powerful both in the number and character of its members, and a society was formed as early as 1701, undoubtedly the pioneer of this creed in New Hampshire, as an organized and healthy society, and the descendants of the members of it having emigrated to other towns became the nucleus or prominent support of other societies, and their names reappear, as the Felches of Tamworth, the Goves of Weare, etc. The first Presbyterian Society was organized in 1764. Seabrook was formed of the territory reclaimed from Massachusetts and a portion of old Hampton, and derived its name from having the sea on one side, into which it discharges numerous large brooks. The building of whale and other boats is more extensively carried on than in any other town in the State. Dearborn Academy was founded in 1851, and a substantial brick edifice was erected two years later. An endowment of fifteen thousand dollars was made by the late Dr. Edward Dearborn, an eminent physician and distinguished citizen. [He also left a fund of four thousand dollars, the proceeds of which are "to be applied to the support of the gospel forever in this place."] It has a pleasant and salubrious situation in the village, commanding extensive views of neighboring villages, distant mountains, and the "deep blue sea."

The early settlers of this, as most other towns, suffered much from the depredations of Indians, and on one occasion a band, numbering thirty-two, by the count of a Mr. Gove, who watched them as they crawled on their hands and knees from the swamp, visited the town, and first killed a widow Henry, beating out her brains with tomahawks. She was much lamented by the society of Friends, she being one of their most effective preachers. An earthen vessel, she was carrying at the time she was attacked and killed, is still in the possession of Jonathan Gove. Thomas Lancaster, who was on his way to mill, was the next person killed. Jonathan Green was next killed, being bruised and beaten in a cruel and horrible manner. A small child was next seized, and its head mangled by striking it against a plough. They then entered the house of Nicholas Bond, killed and scalped him, and, having perpetrated all the mischief they could without too much exposure, made their escape.

Seabrook is considered a valuable farming town, is under a high state of cultivation, has extensive salt marshes, large forests of pine wood and timber, and the Eastern Railroad passes nearly through the centre of the town, affording easy and rapid communication with Boston, as well as the north and east. It joins Salisbury on the east, and is a near neighbor to Newburyport.

It will be seen that the Merrimack River has its source in the Willey and neighboring mountains of the majestic group of crystal or white hills in the great ungranted territory of northern New Hampshire, at an altitude of some six thousand feet above the ocean, in the heart of a region so wild and extensive as to be but little known; that it pursues a south-westerly course for the first forty or fifty miles, then south, through the State, thence east to the ocean; that it is some two hundred and sixty miles in length by its course; that its tributaries number twenty-five respectable rivers, and smaller streams innumerable; that all of these of any consequence, except the Winnepesaukee and the Suncook, flow to it from the west, north, or south-west, and all of them except the Concord have their course, wholly or in part, through New Hampshire. It will also be observed that the incorporated capital invested in manufacturing amounts in round numbers to forty million dollars, besides an investment in mills and mechanical works, not incorporated, estimated at one-half as much more, exclusive of the wealth collected and accumulated, represented by property entirely disconnected with and independent of these interests although to a great extent created by them, the improvements, and the vastly increased value of every species of property throughout its whole course; while almost wholly by reason of the facilities afforded for manufacturing by the unequalled water-power of the Merrimack, the population along the river has increased from a few slender and scattered hamlets to more than two hundred thousand souls. On the Merrimack River proper there are five heavy, substantial, and expensive dams, and some twenty-five railway and highway bridges. The figures already given will exhibit something of the incredible amount of manufacturing business done on the Merrimack and will, with what has been described as its collateral or undeveloped capacity, give a more definite idea of the hydraulic power of that stream than has heretofore been generally prevalent. More than three-fourths of the population and wealth of the Merrimack is without doubt due solely to the many unequalled water-powers along the course of that stream. The magnitude, complication, and elaboration of the manufacturing business on the Merrimack would astound and bewilder those unacquainted with the production of textile fabrics on a large scale. The most eminent hydraulic engineers in the land have disclosed the superiority of mind over matter, in capturing the resistless current of this magnificent river and impressing it at pleasure into the service of civilization and the useful arts. The existence of the splendid system of waterfalls, such as this alone, of all the streams in the land, can boast, has cited around them mechanics, artisans, and operatives of every degree of skill and ability, and the result is seen in the steady and successful operation of more, than one hundred monster cotton and woollen mills, whose massive walls, towering on the "air line" towards the clouds, enclose gems of humanity as well as of intricate, delicate, and almost intelligent machinery; in the numerous machanic shops the ring and pleasant hum of which is the cheerful and melodious diapason of prosperity; in the springing up with unparalleled rapidity of fine towns and beautiful cities thronged with industrious and intelligent populations. Anterior to the manufacturing epoch it has been seen that though the Merrimack River was the same lovely stream of bright, sparkling water, and contained the same noble falls, and was surrounded with a population sturdy and indomitable, which, sparse and devoted to the pleasant and profitable pursuits of peace as it was, yet contributed its full share to the independence, intellect, and character of the nation. Looking still further back, to the aboriginal period, the Merrimack and the territory which it drains is replete with interest, different in kind to be sure, but equal and in some respects surpassing that which invests it now, and while we condemn the cruelties practised towards the red man and by him, and the wrong and injustice perpetrated upon him in the name of civilization and under the sometimes pliant banner of Christianity, it may be proper even in the plentitude of intelligence, enlightenment, and power, to bestow some thought on his checkered history that we may profit by his example and if possible avoid his follies, his misfortunes, and his deplorable fate. It has been said that the result of the practical operation of what is not inaptly called the philosophy of fate is in direct inverse ratio to rational probabilities; thus, a man may fail to obtain that which he merits, while another may fail to merit that which he obtains; and so also with a community, intelligence may be its corner-stone, good government, the arts and sciences, industry and thrift may give the structure strength, beauty, and durability, power may be the keystone, and apparent permanence one of its chief elements; but who shall say when a race, mentally, morally, or physically superior shall appear, and prostrating all these emblems of stability, might, and perpetuity, effect a change as radical and complete, perhaps as beneficent, as that which has occurred in the comparatively brief period of two hundred years past. But though races of men may flourish for a season and disappear, others more or less worthy assuming their places in turn, the Merrimack River and its grand surroundings can never be involved in these vicissitudes. The grand convocation of majestic mountains which surround its source are the fitting emblems of eternal duration and nothing but such terrific convulsions of nature as would produce a universal chaos could move them from their firm bases, or mar the unequalled natural beauty of their scenery, or destroy the wonderful features which give them a world-wide fame. The Merrimack itself, enduring as these crystal hills which give it birth, will also, through all the changing scenes of this world, still roll on, a feature of great beauty to the land, and a source of perpetual and untold wealth, convenience, and usefulness to the people. The Merrimack River will go on forever, leaping from the great mountains, where it has its origin, in sparkling cascades, meandering through long, shaded avenues of perennial forests, winding its tortuous course around the bases of eternal hills, a robust, rapid river; fretting its banks through extensive sections of cleared and cultivated fields, and rich alluvial intervals, tumbling over grand falls with a mighty roar, and sweeping through the lowlands until in the bosom of the great deep it finds repose. In ages yet to come, when other hands shall direct its power in the artificial channels of usefulness to mankind, other eyes shall see its marvellous beauty, and other tongues relate its story. In another age new and improved monuments may be reared, still testifying to its service and its power, long after the chains which now bind it to the wheels of monster cotton mills are rusted and decayed and become relics of the past, or the antiquarian may rescue from the debris of its present glory vestiges of the history of its former, but fallen, grandeur.

> "By thirty hills I hurry down,
> Or slip between the ridges,
> By twenty thorps, a little town
> And half a hundred bridges —"

and thus singing as it rolls along Tennyson's beautiful song of the brook, — by the eventful history of this most historic place, and the mystery which still surrounds an aboriginal and most mysterious race, "now gone, all gone,"— by the decline and fall of a wilderness empire and its reoccupation by another race and color which, under the ample folds of the broad, bright banner of civilization, has uprooted Paganism, and disseminated a progressive enlightened and Christian faith,— by the going and the coming of races of men which it has witnessed, — the Merrimack gives peculiar emphasis and force to the declaration, —

> ' "For men may come and men may go,
> But I go on forever."

www.ingramcontent.com/pod-product-compliance
Lightning Source LLC
Chambersburg PA
CBHW022106230426
43672CB00008B/1294